THE DESIGN OF LOW NOISE OSCILLATORS

THE DESIGN OF LOW NOISE OSCILLATORS

Ali Hajimiri
California Institute of Technology

Thomas H. Lee
Stanford University

KLUWER ACADEMIC PUBLISHERS
Boston / Dordrecht / London

Distributors for North, Central and South America:
Kluwer Academic Publishers
101 Philip Drive
Assinippi Park
Norwell, Massachusetts 02061 USA
Telephone (781) 871-6600
Fax (781) 871-6528
E-Mail <kluwer@wkap.com>

Distributors for all other countries:
Kluwer Academic Publishers Group
Distribution Centre
Post Office Box 322
3300 AH Dordrecht, THE NETHERLANDS
Telephone 31 78 6392 392
Fax 31 78 6546 474
E-Mail <orderdept@wkap.nl>

Electronic Services <http://www.wkap.nl>

Library of Congress Cataloging-in-Publication Data

A C.I.P. Catalogue record for this book is available
from the Library of Congress.

Printed on acid-free paper.

Printed in the United States of America

To Our Parents

Contents

Preface

It is hardly a revelation to note that wireless and mobile communications have grown tremendously during the last few years. This growth has placed stringent requirements on channel spacing and, by implication, on the phase noise of oscillators. Compounding the challenge has been a recent drive toward implementations of transceivers in CMOS, whose inferior $1/f$ noise performance has usually been thought to disqualify it from use in all but the lowest-performance oscillators.

Low noise oscillators are also highly desired in the digital world, of course. The continued drive toward higher clock frequencies translates into a demand for ever-decreasing jitter.

Clearly, there is a need for a deep understanding of the fundamental mechanisms governing the process by which device, substrate, and supply noise turn into jitter and phase noise. Existing models generally offer only qualitative insights, however, and it has not always been clear why they are not quantitatively correct.

This monograph offers a new time-variant phase noise model. By discarding the implicit assumption of time-invariance underlying many other approaches, this model is capable of making quantitative predictions of the phase noise and jitter of different types of oscillators. It is able to attribute a definite amount of phase noise to every noise source in the circuit. Because of its time-variant nature, the model also takes into account the effect of cyclostationary noise sources in a natural way. It details the precise mechanism by which low frequency noise, such as $1/f$ noise, upconverts into close-in phase noise. An important new understanding is that rise and fall time symmetry controls such upconversion. More important, it suggests practical methods for suppressing this upconversion, so that good oscillators can be built in technologies with notoriously poor $1/f$ noise performance (such as CMOS or GaAs MESFET).

The time-variant phase noise model reduces to previously published phase noise models as special cases, provided that certain restrictive assumptions (which can now be stated explicitly) are applied. The theory is verified experimentally for a large number of oscillators with different topologies.

Both tuned and relaxation oscillators are subsumed in a single treatment. Of the latter class, ring oscillators are of particular interest because they perform clock generation in many applications. An expression for the phase noise of ring oscillators in terms of power dissipation, frequency and other circuit parameters is obtained using an approximate analytical model. Among the insights offered, it is shown that the optimum number of stages resulting in minimum phase noise for a given power dissipation and frequency differs for single-ended and differential ring oscillators.

In addition, the theory also accommodates correlations among noise sources. It is shown that it is possible to exploit the strong correlation among the supply and substrate noise sources to minimize their effect on jitter. The approach is generalized to multiple noise sources with arbitrary correlation and cyclostationarity.

This work is divided into seven chapters. The first offers a brief introduction, while Chapter 2 reviews the definition, importance, and modeling of frequency instabilities in electrical oscillators. Chapter 3 summarizes some of the existing models for jitter and phase noise, and Chapter 4 presents the time-variant phase noise model which will be used in the subsequent chapters. Prediction of jitter and phase noise of ring oscillators is the subject of Chapter 5, while the phase noise of differential LC oscillators is investigated in Chapter 6. Finally, generalization of the approach to multiple noise sources is performed in Chapter 7.

The authors are greatly indebted to numerous colleagues among the students and staff at Stanford, as well as at other organizations and institutions, for their many contributions to this monograph.

Ali Hajimiri

Thomas H. Lee

CHAPTER 1 Introduction

In wireless communications, the frequency spectrum is a valuable commodity as the ever increasing number of wireless users demands more efficient usage of the already scarce frequency resources. Communication transceivers rely heavily on frequency conversion using local oscillators (LOs) and therefore the spectral purity of the oscillators in both the receiver and the transmitter is one of the factors limiting the maximum number of available channels and users. For that reason, a deeper understanding of the fundamental issues limiting the performance of oscillators, and development of design guidelines to improve them, are necessary.

During the last fifteen years, there has been tremendous growth in wireless mobile systems [1]-[3]. These systems have been made possible by technological advances in the field of integrated circuits (ICs) allowing a high level of integration at low cost and low power dissipation. There is also great interest in integrating complete communication transceivers on a single chip. This single chip implementation of the systems results in a new environment for oscillators which has not been investigated before.

In digital applications, the timing accuracy of the clock signal determines the maximum clock rate and hence the maximum number of operations per unit time. In microprocessors and other synchronous very large scale digital circuits, the clock signal is generated by on-chip oscillators locked to an external oscillator. Ring oscillators are commonly used for on-chip clock generation due to their large tuning range and ease of integration. In the IC environment, there are additional sources affecting the frequency stability of the oscillators, namely, substrate and supply noise arising from switching in the digital circuitry and output drivers. This new environment and the delay-based nature of ring oscillators demand new approaches to the modeling and analysis of the frequency stability of the oscillators.

It will be shown that all oscillators are periodically time-varying systems, and that their time varying nature must, therefore, be taken into account to permit accurate modeling of phase noise. Also due to the periodically changing operation points of the active devices in the oscillator, many of the noise sources have a periodically time-varying power spectrum; they are cyclostationary. In this work, a time-variant model which is capable of properly assessing the effects of both stationary and cyclostationary noise sources is presented.

The approach presented here explains the exact mechanism by which spurious sources, random or deterministic, are converted into phase and amplitude variations. This time variant model makes explicit predictions about the relationship between waveform shape and $1/f$ noise upconversion. It also shows that the upconversion can be reduced by exploiting the symmetry properties of the waveform. This result is particularly important in CMOS oscillators because it shows that the effect of inferior $1/f$ device noise can be reduced by proper design.

1.1 Organization

Chapter 2 gives an introduction to the definition, importance, and modeling of frequency instabilities in electrical oscillators. It describes some of the different methods for quantifying frequency instability in an oscillator. Some negative consequences of frequency instability in analog, digital, and communication circuits are described in detail.

Chapter 3 reviews some of the existing models for jitter and phase noise in electrical oscillators. It reviews the Leeson model and its extensions for prediction of phase noise in tuned-tank oscillators. It also reviews existing models for jitter and phase noise in ring oscillators.

Chapter 4 presents the core of the time variant phase noise model. It introduces the *impulse sensitivity function* (ISF), from which the phase and amplitude response of an oscillator to an arbitrary noise source may be determined. It elucidates the details of low frequency noise upconversion process and accommodates time variant effects such as cyclostationary noise sources and voltage dependent capacitors. It also investigates modeling of amplitude noise, as opposed to phase noise. Finally, previously existing models are shown as special cases of the general theory.

Prediction of jitter and phase noise of ring oscillators is the subject of Chapter 5. It starts with an approximate method to calculate the dc and rms values of the ISF for ring oscillators with equal and unequal rise and fall times. The rms and dc values are then used to find approximate analytical expressions for the phase noise and jitter of differential and single-ended ring oscillators in terms of the number of stages, power dissipation, frequency, and circuit parameters. The effect of correlated noise sources on the ring oscillator is demonstrated next. Design implications such as the question of single-ended vs. differential implementation of ring oscillators and the optimum number of stages are addressed. Finally, experimental results are compared with predictions made by the theory.

Chapter 6 investigates the phase noise analysis of differential LC oscillators. A simple expression for the tank amplitude is obtained. The effect of different noise sources in such oscillators is analyzed and methods for exploiting the cyclostationary properties of noise are shown. The effect of tail current noise on the phase noise is explored. Finally, new design implications arising from this approach and experimental results are given.

In Chapter 7, the phase impulse response concept introduced in Chapter 4 is generalized to accommodate multiple noise sources. Using the generalized impulse response of the system, the output spectrum of the system in the presence of multiple noise sources with arbitrary correlation and cyclostationarity is calculated.

A summary of the results is given in Chapter 8.

Appendix A gives the relationship between jitter and phase noise. Appendix B shows the calculations leading to the Lorentzian power spectrum of the output. In Appendix C, the ISF for an ideal LC oscillator is directly calculated by solving circuit differential equations. Appendix D presents two methods for calculations of the ISF. Finally, Appendix E investigates jitter and phase noise in phase-locked loops (PLL).

CHAPTER 2

Frequency Instability Fundamentals

Any practical oscillator has fluctuations in its amplitude and frequency. Short term frequency instabilities[1] of an electrical oscillator are mainly due to noise and interference sources. Thermal, shot and flicker noise are examples of the former, while substrate and supply noise are in the latter group. These sources result in frequency instabilities that can be characterized in different ways. Section 2.1 gives an informal introduction to frequency instabilities and their destructive effects on the performance of analog and digital systems. Section 2.2 covers definitions of jitter and phase noise which are the two most commonly used parameters for quantifying frequency instabilities.

2.1 Introduction to Frequency Instability

This section presents a qualitative approach to the meaning and importance of frequency instabilities in both frequency and time domains.

1. Short term frequency instabilities usually refer to variations on time scales smaller than a second. This definition of short term instabilities is arbitrary and does not arise from any fundamental considerations.

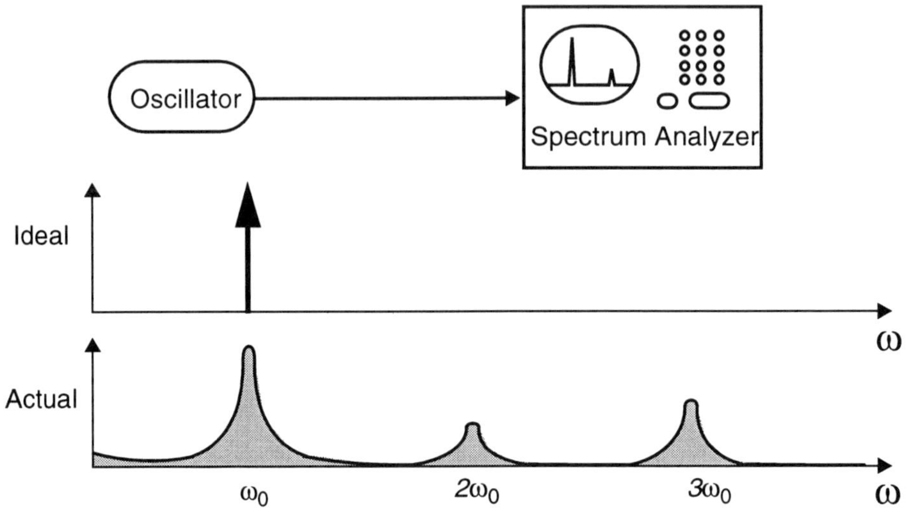

FIGURE 2.1 The spectrum of an ideal and a practical oscillator.

2.1.1 Frequency Domain

The output of an ideal oscillator may be expressed as $V_{out}(t)=V_0 \cos[\omega_0 t+\phi_0]$, where the amplitude V_0, the frequency ω_0, and phase reference ϕ_0 are all constants. The one-sided spectrum of an ideal oscillator with no random fluctuations consists of an impulse at ω_0 as shown in Figure 2.1. In a practical oscillator, however, the output is more generally given by

$$V_{out}(t) = V_0 \cdot [1 + A(t)] \cdot f[\omega_0 t + \phi(t)] \tag{2.1}$$

where $\phi(t)$ and $A(t)$ are functions of time, V_0 is the maximum voltage swing and f is a periodic function which represents the shape of the steady-state output waveform of the oscillator. The output spectrum has power around harmonics of ω_0 if the waveform, f, is not sinusoidal. More important, as a consequence of the fluctuations represented by $\phi(t)$ and $A(t)$, the spectrum of a practical oscillator has sidebands close to the frequency of oscillation, ω_0, and its harmonics, as shown in Figure 2.1. These sidebands are generally referred to as *phase noise* sidebands.

The destructive effect of phase noise can be best seen in the front-end of a superheterodyne radio receiver. Figure 2.2 shows a typical front-end block diagram, consisting of a low noise amplifier (LNA), a mixer, and a local oscillator (LO). Suppose the

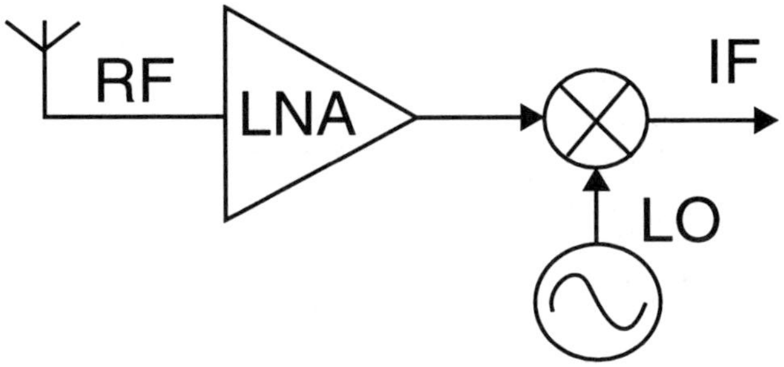

FIGURE 2.2 Simplified model for the front-end of a radio receiver.

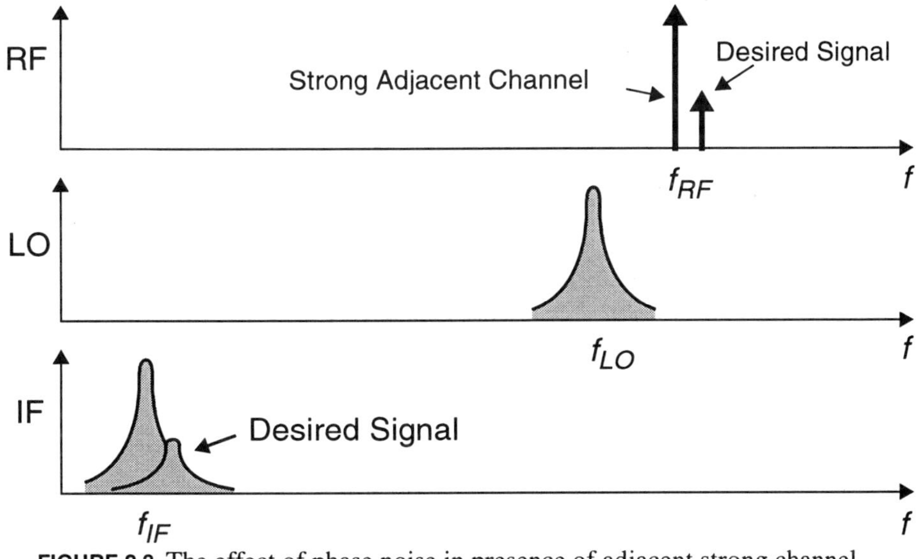

FIGURE 2.3 The effect of phase noise in presence of adjacent strong channel.

receiver tunes to a weak signal in the presence of a strong signal in an adjacent channel. If the LO has large phase noise, as shown in Figure 2.3, some downconversion of the interfering signal into the same IF (intermediate frequency) as that of the desired signal will occur as shown in Figure 2.3. The resulting interference significantly degrades the dynamic range of the receiver. Therefore, improving the phase noise of the oscillator clearly improves the signal-to-noise ratio of the desired signal.

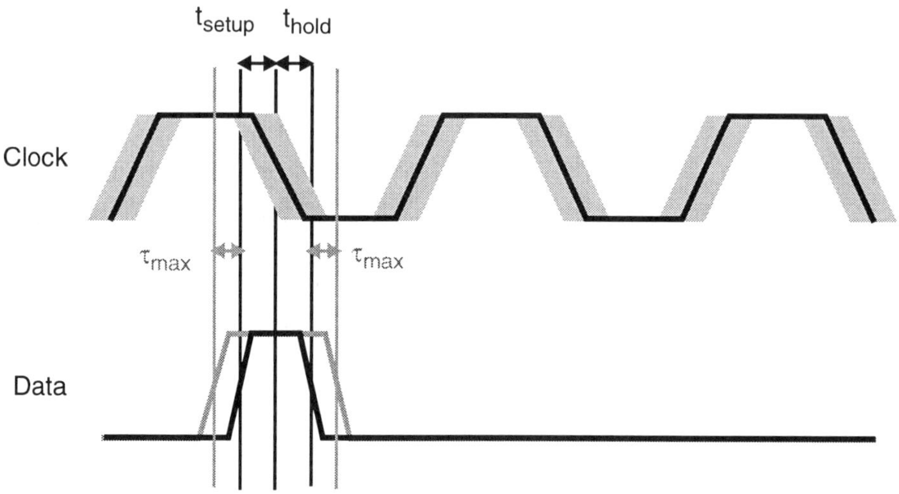

FIGURE 2.4 The effective timing margin in presence of jitter.

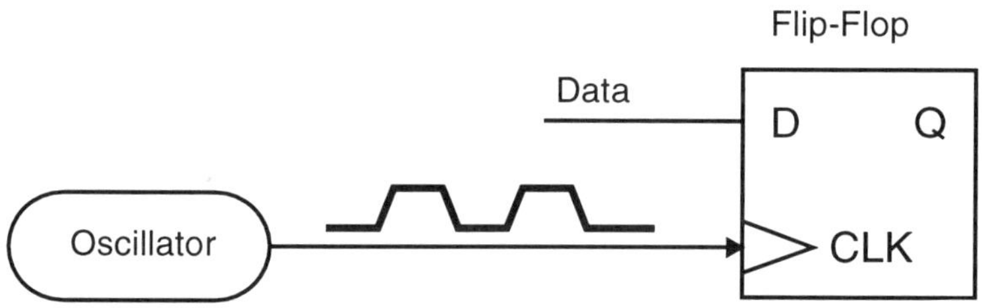

FIGURE 2.5 A flip-flop in a digital circuit.

2.1.2 Time Domain

In the time domain viewpoint, the spacing between transitions is ideally constant. In practice, however, the transition spacings will be variable due to fluctuations in $\phi(t)$. This uncertainty is known as *timing jitter* and can be seen in Figure 2.4.

In a synchronous digital circuit such as a microprocessor, there is a clock signal that controls the operation of different logic blocks. To emphasize the importance of timing jitter, consider the example of a flip-flop shown in Figure 2.5. If the clock signal has zero timing jitter as shown with the solid line in Figure 2.4, the data needs to be stable only for $t_{setup}+t_{hold}$. However, if the clock line shows a peak-to-peak jitter of

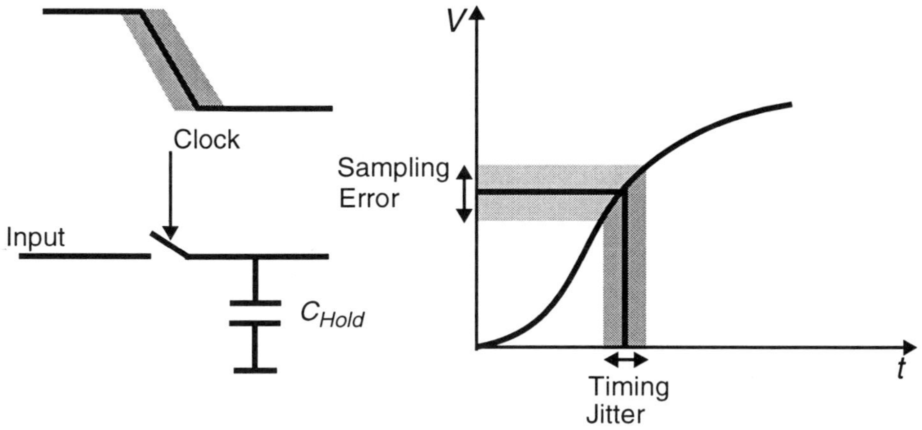

FIGURE 2.6 The effect of timing jitter sampling error.

τ_{max}, then the data line needs to be stable for a period of $t_{setup}+ t_{hold} + 2\tau_{max}$ as shown in Figure 2.4. This decrease in the timing margins will reduce the maximum achievable frequency of operation for the digital circuit.

The harmful effect of clock jitter can also be seen in the sample-and-hold circuit of Figure 2.6, where the accuracy of the sampling process is affected by jitter in the clock. If there is uncertainty in sampling time (*i.e.,* clock jitter), it translates directly to uncertainty in the sampled value (*i.e.,* noise) as shown in Figure 2.6.

2.2 Frequency Instability Characterization

As shown in the last section, there are several ways of quantifying short-term frequency instabilities of an oscillator. While comprehensive reviews of various standards and measurement methods can be found in [23][24] and [57]-[63], the focus of this section will be on the two most popular quantities for characterizing these fluctuations, namely, phase noise and timing jitter. The qualitative treatment of these fluctuation measures will be expanded upon in this section.

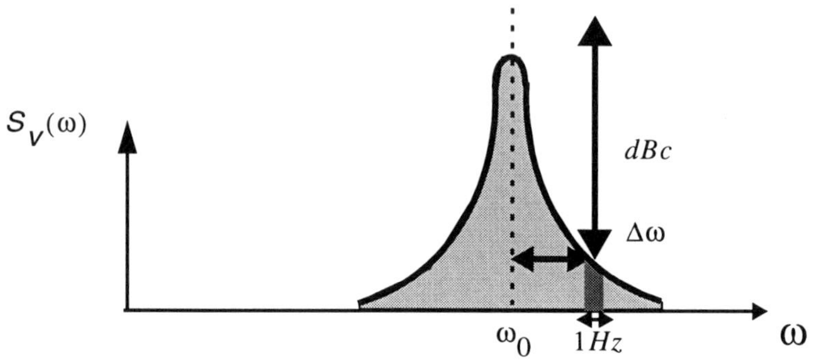

FIGURE 2.7 The phase noise per unit bandwidth.

2.2.1 Phase Noise

In the frequency domain viewpoint, an oscillator's short term instabilities are usually characterized in terms of the single sideband noise spectral density. It is conventionally given the units of decibels below the carrier per Hertz (dBc/Hz) and is defined as:

$$\mathcal{L}_{total}\{\Delta\omega\} = 10 \cdot \log\left[\frac{P_{sideband}(\omega_0 + \Delta\omega, 1Hz)}{P_{carrier}}\right] \quad (2.2)$$

where $P_{sideband}(\omega_0 + \Delta\omega, 1\text{Hz})$ represents the single sideband power at a frequency offset, $\Delta\omega$, from the carrier in a measurement bandwidth of 1Hz as shown in Figure 2.7, and $P_{carrier}$ is the total power under the power spectrum. Note that the definition in (2.2) includes the effect of both amplitude and phase fluctuations, $A(t)$ and $\phi(t)$.

Spectral density is usually specified at one or a few offset frequencies. To be a meaningful parameter, both the noise density and the offset need to be reported, *e.g.*, -121dBc/Hz at 600kHz offset from the carrier.

The advantage of $\mathcal{L}_{total}\{\Delta\omega\}$ in (2.2) is its ease of measurement. Its disadvantage is that it shows the sum of both amplitude and phase variations; it does not show them separately. It is often important to know the amplitude and phase noise separately because they behave differently in a circuit. For instance, the effect of amplitude noise can be reduced by amplitude limiting, while the phase noise cannot be reduced in an analogous manner. Therefore, in most practical oscillators, $\mathcal{L}_{total}\{\Delta\omega\}$ is dominated

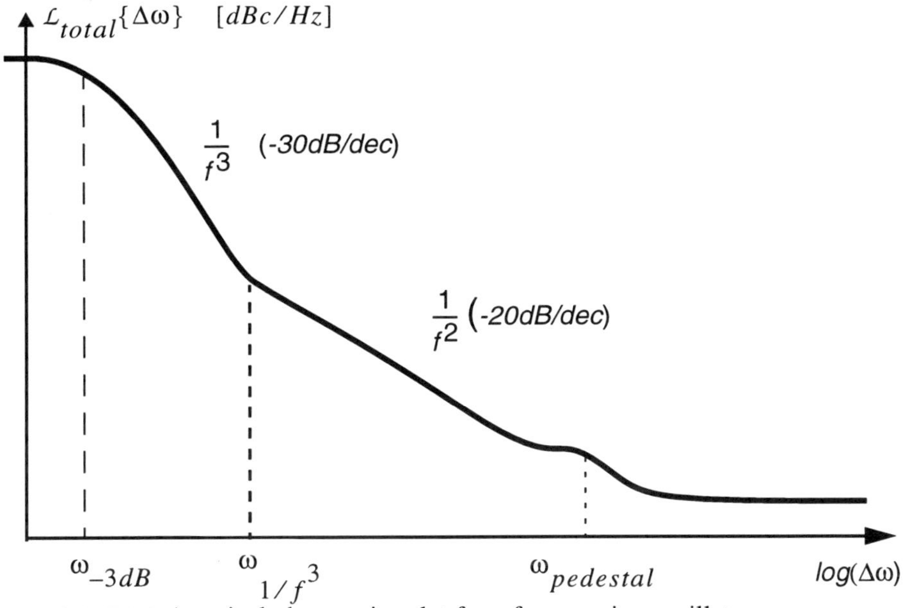

FIGURE 2.8 A typical phase noise plot for a free running oscillator.

by its phase portion, $\mathcal{L}_{phase}\{\Delta\omega\}$, known as phase noise, which will be simply denoted as $\mathcal{L}\{\Delta\omega\}$, unless specified otherwise.

If one plots $\mathcal{L}_{total}\{\Delta\omega\}$ for a free-running[1] oscillator as a function of $\Delta\omega$ on logarithmic scales, regions with different slopes may be observed as shown in Figure 2.8. At large offset frequencies, there is a flat noise floor[2]. At small offsets, one may identify regions with a slope of $1/f^2$ and $1/f^3$, where the corner between the $1/f^2$ and $1/f^3$ regions is called ω_{1/f^3}. Finally, the spectrum becomes flat again at very small offset frequencies. The mechanisms responsible for these features will be discussed in great detail in subsequent chapters.

1. An oscillator which is not locked to a reference is referred to as free running. The phase noise behavior of an oscillator in a feedback loop is the subject of Appendix E.

2. A bump, such as the feature shown preceding the noise floor in Figure 2.8, may also result from the effect of amplitude fluctuations. This phenomenon will be discussed in more detail in Section 4.6.

There are different methods of measuring phase noise and, depending on the particular method used to measure it, parts of the spectrum in Figure 2.8 may or may not be observed. For example, if a spectrum analyzer is used to measure the phase noise, ω_{-3dB} will be easily observed. However, if the phase noise is measured using a phase-locked loop, the nonlinear transfer function of the phase detector[1] will change the measured ω_{-3dB}. A very complete review of these measurement techniques and their properties can be found in [57]-[63].

It is instructive to calculate the required phase noise specifications for the local oscillator in the example of Figure 2.2. In particular, the phase noise required to achieve a desired signal-to-noise ratio can be estimated for a given interfering signal power using the definition of (2.2). This estimation can be performed by calculating the total inband noise with respect to the carrier. The inband noise relative to the carrier is found by integrating the phase noise spectrum over the band of interest, *i.e.*,

$$P_{noise} = \int_{\Delta f_{min}}^{\Delta f_{max}} \mathcal{L}\{\Delta f\} d(\Delta f) \quad \textbf{(2.3)}$$

where Δf_{min} and Δf_{max} are the offsets from the center of the channel to the edges of the adjacent channel.

Assuming that the phase noise has a $1/f^2$ slope between Δf_{min} and Δf_{max}, (2.3) reduces to

$$P_{noise} = \mathcal{L}\{\sqrt{\Delta f_{min} \cdot \Delta f_{max}}\} \cdot (\Delta f_{max} - \Delta f_{min}) \quad \textbf{(2.4)}$$

which means it is equivalent to the noise due to a constant phase noise of $\mathcal{L}\{\sqrt{\Delta f_{min} \cdot \Delta f_{max}}\}$ between Δf_{min} and Δf_{max}.

The minimum *SNR* (signal-to-noise ratio) is given by

$$10\log(SNR_{min}) = 10\log(P_{sig}) - [10\log(P_{int}) + 10\log(P_{noise})] \quad \textbf{(2.5)}$$

where P_{sig} and P_{int} are the desired and interfering signal powers, respectively.

For example, assume a channel spacing of 200kHz, so that Δf_{min}=100kHz and Δf_{max}=300kHz. If an adjacent interferer is 40dB stronger than the desired signal, the

1. The properties of phase-locked loops and phase detectors are discussed in Appendix E.

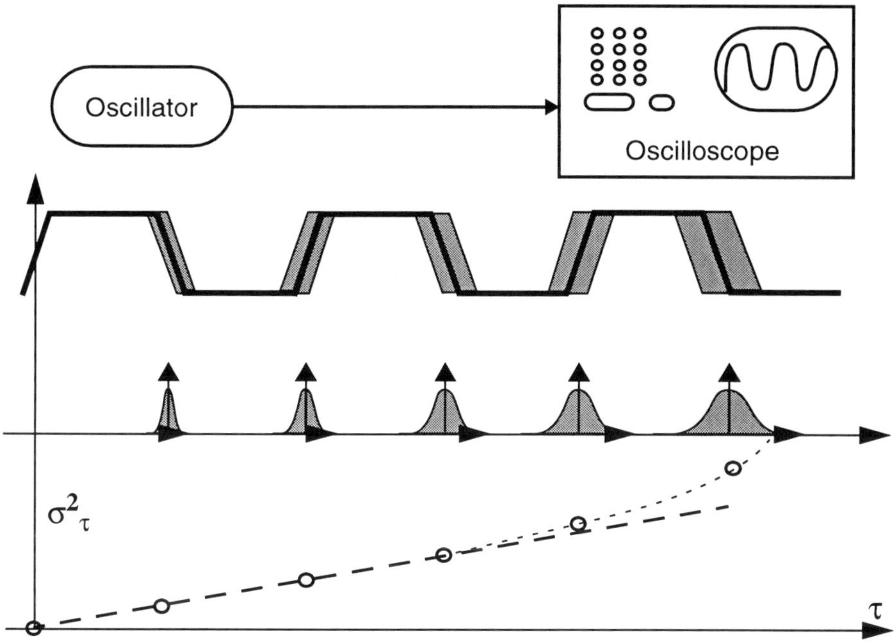

FIGURE 2.9 Clock jitter increasing with distance from the reference edge.

maximum allowable phase noise for a minimum *SNR* of 20dB is calculated from (2.5) to be -113dBc/Hz at a 173kHz offset from the carrier. Assuming a $1/f^2$ slope, this specification is equivalent to a phase noise of -108dBc/Hz at a 100kHz offset.

-113 dBc/Hz over 200 KHz ⇒ -60 dBc $\log_{10}(2e^5) = 53$

As can be seen from the foregoing calculation, only small values of phase noise are permitted, even for modest values of *SNR* and required adjacent blocking channel power. The requirements become even more severe as the carrier frequency increases and channel spacing shrinks.

2.2.2 Timing Jitter

As mentioned earlier, uncertainties in the transition instants of a periodic waveform are known as clock jitter. For a free-running oscillator, it increases with the measurement interval τ (*i.e.,* the time delay between the reference and the observed transitions). This increase is illustrated in the plot of timing variance shown in Figure 2.9 [48].

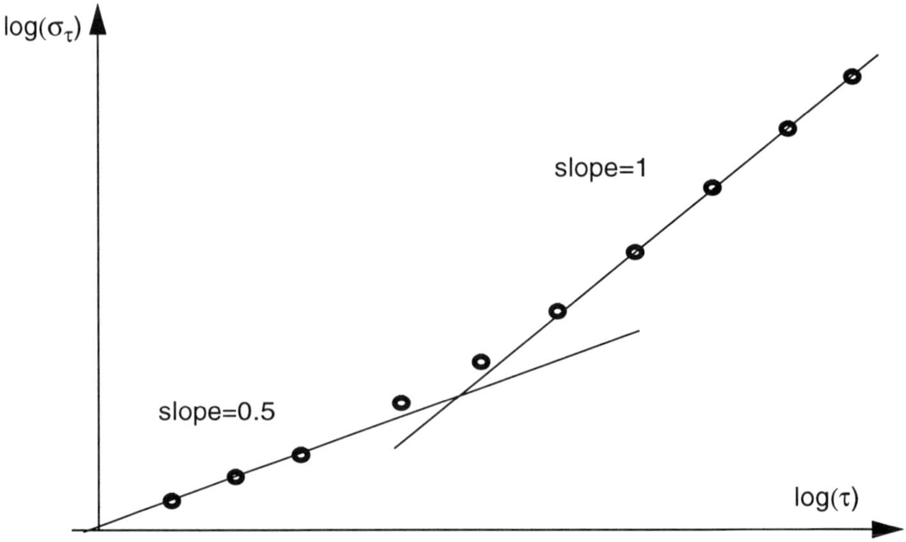

FIGURE 2.10 Rms jitter vs. measurement time on a log-log plot.

This growth in variance (*i.e.*, "jitter accumulation") occurs because any uncertainty in an earlier transition affects all the following transitions, and its effect persists indefinitely. Therefore, the timing uncertainty when τ seconds have elapsed includes the accumulative effect of the uncertainties associated with the transitions.

A *log-log* plot of the timing jitter, σ_τ, versus the measurement delay, τ, for a free-running oscillator will typically exhibit regions with slopes of $\frac{1}{2}$ and 1 as shown in Figure 2.10. In the region with the slope of $\frac{1}{2}$, the standard deviation of the jitter after τ seconds is

$$\sigma_\tau = \kappa\sqrt{\tau} \tag{2.6}$$

where κ is a proportionality constant determined by circuit parameters. In a similar fashion, the standard deviation of the jitter in the region with the slope of 1 may be expressed as

$$\sigma_\tau = \zeta \cdot \tau \tag{2.7}$$

where ζ is another proportionality constant.

In most digital applications, it is desirable for σ_τ to decrease at the same rate as the frequency increases, to keep constant the ratio of the rms timing jitter to the period. Therefore, phase jitter, defined as

$$\sigma_{\Delta\phi} = 2\pi\frac{\sigma_\tau}{T} = \omega_0\sigma_\tau, \tag{2.8}$$

is a more useful measure in many applications. In (2.8), T is the period of oscillation.

The definition in (2.6) and (2.7) can be used to estimate the tolerable timing jitter for a required *SNR* in a sample-and-hold circuit. Specifically, consider applying a sinusoidal input voltage, $V_0\sin(\omega_0 t)$, to the sample and hold circuit of Figure 2.6. If the sampling clock applied to an ideal switch has timing jitter with a standard deviation of σ_τ, the equivalent error in the sampled voltage, σ_V, will be related to the timing jitter through the slope of the sinusoid, *i.e.*,

$$\sigma_V = V_0\omega_0\cos(\omega_0 t)\sigma_\tau \tag{2.9}$$

If there is no correlation between the sampling clock and the input waveform (usually a reasonable assumption), the signal-to-noise ratio can be calculated by averaging the power of the voltage error in (2.9). Thus, if the signal-to-noise ratio (*SNR*) is limited by jitter, it will be given by [143],

$$SNR = \frac{V_0^2/2}{\sigma_{V,\,ave}^2} = \frac{V_0^2/2}{V_0^2\omega_0^2\sigma_\tau^2/2} = \frac{1}{\omega_0^2\sigma_\tau^2} \tag{2.10}$$

where $\sigma_{V,\,ave}$ is the average standard deviation of the equivalent voltage noise. A numerical example can be enlightening. If an *SNR* of 60dB is required at a sampling frequency of 10MHz, (2.10) predicts that the rms jitter must not exceed 16psec.

The sensitivity to the sampling time becomes even more important in a *subsampling system*. Here the sampling uncertainty should be compared with the period of the high frequency input signal, not the subsampling period, because the jitter spectral density increases as the square of the subsampling ratio [121].

Assume that this sampling is performed in a communications receiver, in which signals in an adjacent channel $\Delta\omega$ away should be rejected. In this case, σ_τ will correspond approximately to the rms jitter between two transitions $\tau = 2\pi/\Delta\omega$ apart from each other. τ is the time it takes for two sinusoids with frequencies $\Delta\omega$ apart to

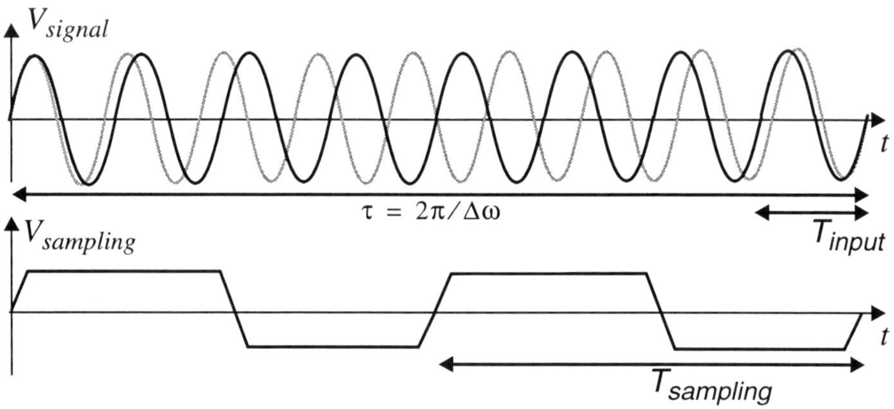

FIGURE 2.11 The time needed for two sinusiodal voltages to mismatch by one cycle.

mismatch by exactly one cycle as shown in Figure 2.11. Assuming a square-root dependence on τ, (2.6) will result in

$$\frac{\sigma_{CTC}}{\sigma_\tau} = \sqrt{\frac{T_{sampling}}{\tau}} = \sqrt{\frac{\Delta\omega}{\omega_s}} \tag{2.11}$$

where σ_{CTC} is the cycle-to-cycle jitter of the sampling clock and $T_{sampling}$ and ω_s are sampling clock period and angular frequency, respectively. Using (2.10), the following expression for the maximum tolerable cycle-to-cycle jitter is obtained

$$\sigma_{CTC} = \frac{1}{\omega_0}\sqrt{\frac{\Delta\omega}{\omega_s \cdot SNR}} \tag{2.12}$$

where *SNR* is the desired signal-to-noise ratio. Note that in a subsampling system, ω_0 is the angular frequency of the incoming high frequency signal and not the subsampling frequency.

For an input frequency of 900MHz, a sampling frequency of 90MHz (corresponding to a subsampling ratio of 10) and a desired *SNR* of 60dB at the center of an adjacent channel 200kHz away, (2.12) permits a maximum rms cycle-to-cycle jitter of 8.3fsec. It is exceedingly difficult to achieve such small values in practice.

CHAPTER 3

Review of Existing Models

The modeling of frequency instabilities has been the subject of numerous studies in different disciplines [13]-[50]. This chapter presents a brief review of some of this earlier work. Section 3.1 reviews the Leeson model [22] and its extensions [25][37] for tuned-tank oscillators, while Section 3.2 reviews existing models [45]-[48] for ring oscillators.

3.1 Tuned-Tank Oscillators

The phase noise model proposed in [22] and later expanded in [25][37] is widely known as the Leeson model, and is by far the most well-known. It is based on a linear time-invariant (LTI) approach for tuned tank oscillators. It predicts the following behavior for $\mathcal{L}\{\Delta\omega\}$:

$$\mathcal{L}\{\Delta\omega\} = 10 \cdot \log\left[\frac{2FkT}{P_s} \cdot \left[1 + \left(\frac{\omega_0}{2Q_L\Delta\omega}\right)^2\right] \cdot \left(1 + \frac{\omega_{1/f^3}}{|\Delta\omega|}\right)\right] \tag{3.1}$$

where F is an empirical parameter (often called the "device excess noise number"), k is Boltzmann's constant, T is the absolute temperature, P_s is the average power dissipated in the resistive part of the tank, ω_0 is the oscillation frequency, Q_L is the effective quality factor of the tank with all loadings accounted for (also known as loaded

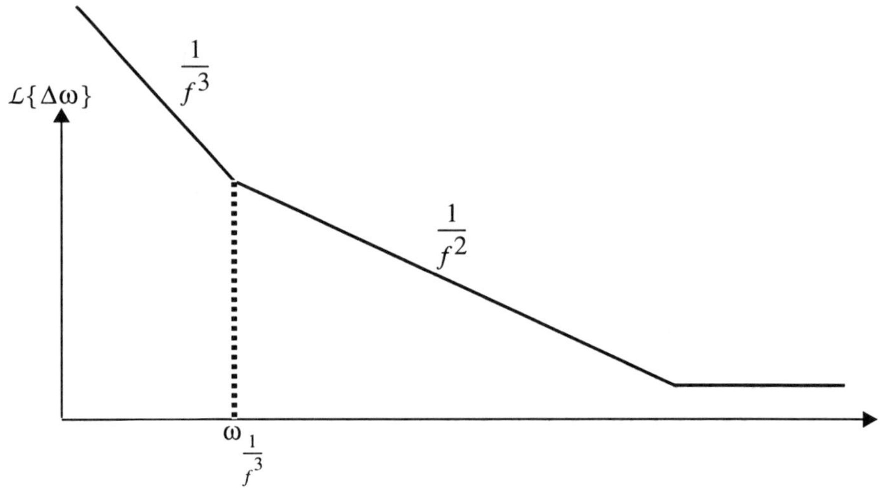

FIGURE 3.1 The spectrum of the phase noise.

Q), $\Delta\omega$ is the offset from the carrier, and ω_{1/f^3} is the frequency of the corner between the $1/f^3$ and $1/f^2$ regions, as shown in Figure 3.1.

The existence of a $1/f^2$ region can be anticipated by applying an LTI approach as follows. The impedance of a parallel RLC tank, for $\Delta\omega \ll \omega_0$, is easily calculated to be

$$Z(\omega_0 + \Delta\omega) \approx \frac{1}{G_L} \cdot \frac{1}{1 + j2Q_L\frac{\Delta\omega}{\omega_0}} \tag{3.2}$$

where G_L is the parallel parasitic conductance of the tank.

To sustain oscillations, the average energy provided to the tank by the active device should be equal to the energy losses in the resonant circuit[1]. Therefore, the active device can be modeled as an effective parallel negative conductance, $-G_m(V_0)$, whose value depends on the tank amplitude, V_0. For steady-state oscillation, the equation $G_m(V_0) = G_L$ should be satisfied. When this condition holds, the net impedance of the oscillator model shown in Figure 3.2 is given by

1. This viewpoint is often referred to as the one-port approach.

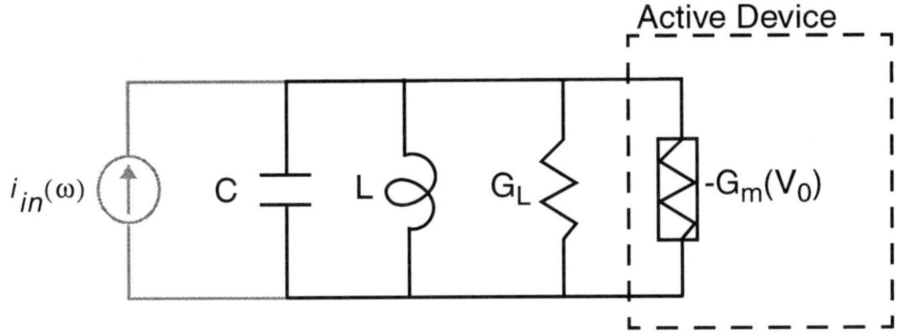

FIGURE 3.2 Equivalent one-port circuit for an LC oscillator.

$$Z(\Delta\omega) = \frac{v_{out}(\omega_0 + \Delta\omega)}{i_{in}(\omega_0 + \Delta\omega)} = -j\frac{1}{G_L} \cdot \frac{\omega_0}{2Q_L\Delta\omega} \tag{3.3}$$

The total equivalent parallel resistance of the tank has an equivalent mean square noise current density of $\overline{i_n^2}/\Delta f = 4kTG_L$. In addition, active device noise usually contributes a significant portion of the total noise in the oscillator. It is traditional to combine all the noise sources into one effective noise source, expressed in terms of the resistor noise with a multiplicative factor, *F*, known as the device excess noise number. The equivalent mean square noise current density can therefore be expressed as $\overline{i_n^2}/\Delta f = 4FkTG_L$. Unfortunately, it is generally difficult to calculate *F a priori*. One important reason, to be shown later, is that much of the noise in a practical oscillator arises from periodically time-varying processes which are not properly treated in an LTI context. Hence, as mentioned in [22], *F* is usually an *a posteriori* fitting parameter derived from measured data.

Using the effective noise current power, the phase noise in the $1/f^2$ region of the spectrum can be calculated as

$$\mathcal{L}\{\Delta\omega\} = 10 \cdot \log\left(\frac{\overline{v_{noise}^2}}{\overline{v_{sig}^2}}\right) = 10 \cdot \log\left(\frac{\frac{1}{2} \cdot |Z(\Delta\omega)|^2 \cdot \overline{i_n^2}/\Delta f}{\frac{1}{2} \cdot V_0^2}\right)$$
$$= 10 \cdot \log\left[\frac{2FkT}{P_s} \cdot \left(\frac{\omega_0}{2Q\Delta\omega}\right)^2\right] \tag{3.4}$$

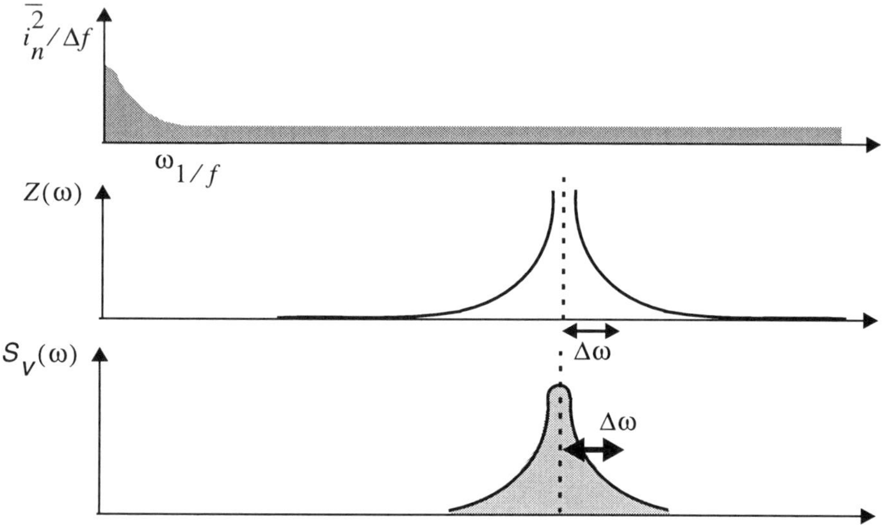

FIGURE 3.3 Conversion of noise current to phase noise sidebands.

This process by which the noise current becomes phase noise sidebands is shown in the frequency domain picture of Figure 3.3. The factor of 1/2 in the numerator of (3.4) arises from neglecting the contribution of amplitude noise, which is here assumed to be suppressed by a suitable amplitude limiting mechanism [22].

Although the phase noise in the $1/f^2$ region of (3.1) is thus easily calculated in (3.4), a similar treatment does not lead to an expression for the $1/f^3$ portion of the phase noise in the empirical expression (3.1), as can be seen from Figure 3.3. Hence, ω_{1/f^3} is just another fitting parameter.

The foregoing approach has been extended by accounting for the individual noise sources in the tuned tank oscillator model of Figure 3.4 [37]. Unfortunately, the approach taken in [37] continues to assume linear time-variance, and adds an unsupported implicit assumption of no amplitude limiting. For the circuit of Figure 3.4, this approach predicts the following:

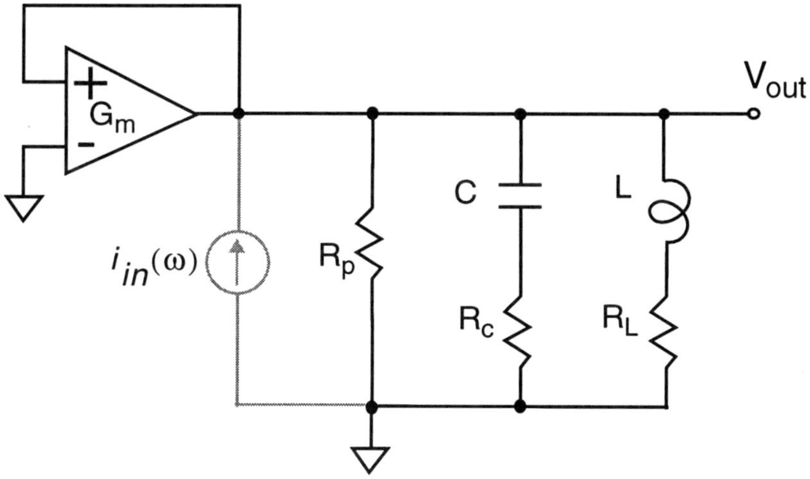

FIGURE 3.4 Equivalent circuit for the LC oscillator of [37].

$$\mathcal{L}\{\Delta\omega\} = 10 \cdot \log\left[\frac{kT \cdot R_{eff}[1 + A] \cdot \left(\frac{\omega_0}{\Delta\omega}\right)^2}{V_0^2/2}\right] \tag{3.5}$$

where A is yet another empirical fitting parameter, and R_{eff} is the effective series resistance, given by

$$R_{eff} = R_L + R_C + \frac{1}{R_p(C\omega_0)^2} \tag{3.6}$$

where R_L, R_C, R_p and C are shown in Figure 3.4. Note that it is still not clear how to calculate A from circuit parameters. While this approach is valuable in identifying the relative contribution of each noise source, it represents no fundamental improvement over the methods outlined in [22] and [25].

An alternative view yields additional insight into the Leeson model [22][25]. In this approach, the oscillator is modeled as a two port system, in contrast with the one-port view presented earlier. Figure 3.5 shows a simplified two-port model (still LTI) for an

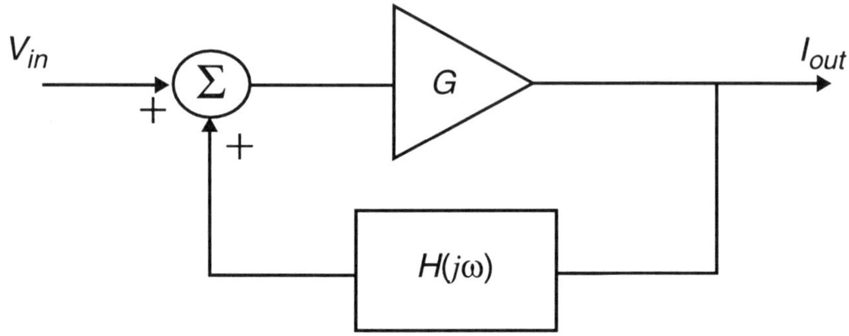

FIGURE 3.5 Simplified two-port model of an oscillator.

oscillator. It consists of a gain block G in the forward path and a frequency selective element with a linear transfer function $H(j\omega)$ in the feedback path.

The transfer function for this linear system is

$$\frac{I_{out}(j\omega)}{V_{in}(j\omega)} = \frac{G}{1 - G \cdot H(j\omega)} \tag{3.7}$$

where V_{in} is the input voltage and I_{out} is the output current. The only way that the feedback system of Figure 3.5 can have non-zero output without any input is for the denominator of (3.7) to be zero, *i.e.*,

$$G \cdot H(j\omega) = 1 \tag{3.1}$$

Thus (3.1), sometimes referred to as the Barkhausen criterion, is a necessary condition for existence of stable oscillations.[1]

In a tuned-tank oscillator, the frequency selective block, $H(j\omega)$, usually consists of an RLC network similar to the one in Figure 3.6. In such a second order system, Q is related to the normalized slope of the phase transfer function,

1. In a purely linear approach the amplitude depends on initial conditions. In practice, excess phase coupled with nonlinearity in the gain block G are used to produce stable-amplitude oscillations.

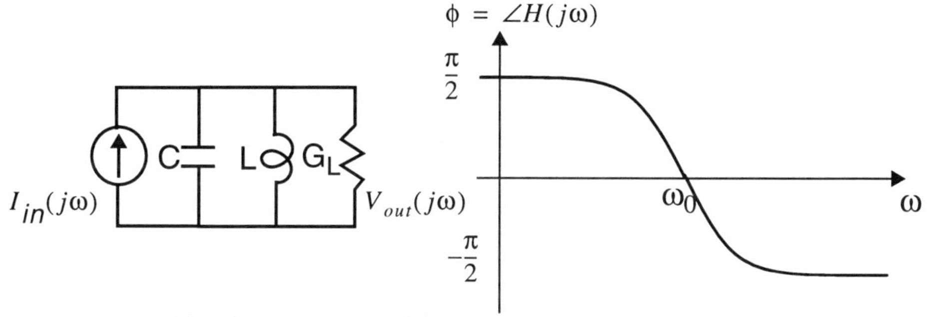

FIGURE 3.6 The phase response of the RLC network.

$$Q = \frac{\omega_0}{2}\left|\frac{d\phi}{d\omega}\right| \tag{3.2}$$

As can be seen from Figure 3.6, a larger Q corresponds to a larger slope in the phase vs. frequency transfer function.

According to (3.1), the total phase shift around the loop has to be an integer multiple of 2π to sustain oscillations. Due to various noise sources in the circuit, temporary phase leads or lags may be introduced in the feedback loop. These extra phase shifts must be compensated by a change in instantaneous frequency according to Figure 3.6 and (3.2). In an oscillator with a large Q, the required instantaneous change in frequency for a given phase shift is smaller, thus resulting in better frequency stability. This observation is in accord with the appearance of the tank's loaded Q in the denominator of (3.4).

In practice, the gain block G will also introduce some phase shift due to the frequency-dependent dynamics of the active devices used to realize it. The oscillation frequency must shift in order to keep the phase around the loop an integer multiple of 2π. This new center frequency will not be where the phase vs. frequency function has the highest slope and hence a larger phase noise will result.

Although this viewpoint leads to some design insight, it is not clear how to adjust this phase shift to gain the best phase noise. Also it is not clear how this approach can be used to improve the phase noise of other types of oscillator, such as ring oscillators. A more general case of this criterion will be discussed in the context of cyclostationary noise sources in Section 4.5.1.

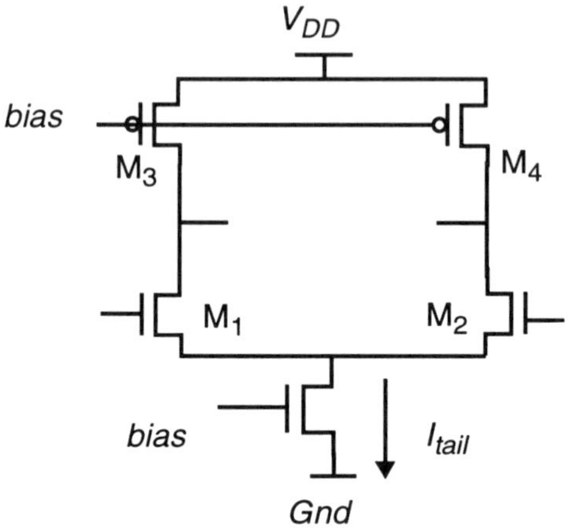

FIGURE 3.7 Differential buffer stages in the ring oscillators of [46].

3.2 Ring Oscillators

A time domain approach to the analysis of differential CMOS ring oscillators is presented in [46]. There, the stage delay is defined as the time interval between zero-crossings of the input and output differential voltages. In a differential ring oscillator with the buffer stages shown in Figure 3.7, the stage delay is approximately

$$t_d = V_{swing}\frac{C_{node}}{I_{tail}} \tag{3.3}$$

where V_{swing} is the single-ended swing on the drain, C_{node} is the total capacitance on the stage output node (including the input capacitance of the next stage), and I_{tail} is the tail current. Using the first-crossing approximation proposed in [26], the standard deviation of the timing jitter due to a single stage, $\sigma_{\tau,1}$, is related to the voltage standard deviation, σ_{vn}, through the maximum transition slope, *i.e.*,

$$\sigma_{\tau,1}^2 = \sigma_{vn}^2\left(\frac{C_{node}}{I_{tail}}\right)^2 \tag{3.4}$$

Assuming that the voltage noise variance is *kT/C* [46], the single stage jitter, normalized to the stage delay, is

$$\frac{\sigma_{\tau,1}^2}{t_d^2} \approx \frac{\sigma_{vn}^2}{V_{swing}^2} \approx \frac{kT}{C_{node}} \cdot \frac{\xi^2}{(V_{GS} - V_T)^2} \tag{3.5}$$

where V_{GS}-V_T is the gate-source overdrive of the differential pair in the balanced state, and ξ is a factor to account for the different shares of the differential pair and load transistors. Therefore, ξ is a topology-dependent parameter.

The cycle-to-cycle jitter can be calculated from the stage jitter of (3.5). For a ring oscillator with N identical stages, assuming independent noise sources, the cycle-to-cycle jitter due to all the stages is given by the sum of variances [46], *i.e.*,

$$\frac{\sigma_{\tau,CTC}^2}{T^2} = \sigma_{\tau,1}^2 \cdot \frac{T}{t_d} = \frac{kT}{I_{tail}} \cdot \frac{a_v \xi^2}{(V_{GS} - V_T)^2} T \tag{3.6}$$

where a_v is the small signal gain of the stage, T is the period of oscillation and $\sigma_{\tau,CTC}$ is the cycle-to-cycle jitter of the output. Equation (3.6) corresponds to a κ, defined in (2.6), of

$$\kappa = \sqrt{\frac{a_v kT}{I_{tail}}} \cdot \frac{\xi}{(V_{GS} - V_T)} \tag{3.7}$$

The phase noise spectrum may also be calculated from (3.6) using the method outlined in [46]:

$$\mathcal{L}\{\Delta\omega\} = 10 \cdot \log\left[\frac{a_v \xi^2 kT}{I_{tail}(V_{GS} - V_T)} \left(\frac{\omega_0}{\Delta\omega}\right)^2\right] \tag{3.8}$$

which shows the trade-offs among power dissipation, number of stages, frequency, and gate overdrive for long channel, differential CMOS ring oscillators.

Since ξ is topology dependent it is not clear how it can be evaluated for an arbitrary oscillator. Also equation (3.8) loses its validity in the presence of short channel effects. Furthermore (3.8) is not valid for inverter-chain or bipolar emitter coupled logic (ECL) ring oscillators. Thus the predictive power of (3.8) is somewhat limited.

Still another time domain approach to predict the jitter of a ring oscillator is presented in [45] and [48]. As in [46], the jitter introduced by each stage of the ring oscillator is assumed to be totally independent of the jitter introduced by the other stages. Therefore, the total variance of the jitter is given by the sum of the variances introduced by each stage. For ring oscillators with identical stages, the variance will be given by $m\sigma_s^2$, where m is the number of transitions during τ, and σ_s^2 is the variance of the uncertainty introduced by one stage during one transition. Noting that m is proportional to τ, (2.6) is obtained.

It is observed in [48] that the κ coefficient in (2.6) for a ring oscillator is independent of the frequency of oscillation and depends on the topology of the inverter stages used. Values of κ due to various noise sources in an emitter coupled bipolar inverter are calculated in [48]. Although this approach thus provides a method for evaluation of the jitter in oscillators, κ needs to be recalculated for each topology and technology.

The foregoing approaches assume statistical independence of noise sources. In reality, however, the statistics of the timing jitter depend on the correlations among the noise sources involved. As an instructive special case, consider totally correlated noise sources. If the timing uncertainties introduced by each stage are fully correlated, their standard deviations, rather than their variances, add. Accordingly, the standard deviation of the jitter after τ seconds is proportional to τ, which can be described by (2.7). A partially correlated source can be broken down into fully correlated and uncorrelated parts. Therefore a *log-log* plot of the timing jitter, σ_τ, versus the measurement delay of τ for a free-running oscillator demonstrates regions with slopes of $\frac{1}{2}$ and 1 corresponding to the uncorrelated and fully-correlated parts of the noise as shown in Figure 2.10.

A frequency domain LTI approach can be used to model the phase noise in differential ring oscillators with a small number of stages, as well as the phase noise of relaxation oscillators [47]. In this approach, a short differential oscillator, such as the 3-stage ring oscillator of Figure 3.8, is modeled using its equivalent small-signal single-ended counterpart shown in Figure 3.9. It is assumed that stages neither turn off nor their transconductances change dramatically during one cycle of operation. Again applying the Barkhausen criterion, the total phase shift around the loop should be 2π in steady state. Also the loop gain, and hence the gain of each stage of Figure 3.9. should be 1. These two conditions translate to $\omega_0 = \sqrt{3}/RC$ and $G_m R = 2$.

The transfer function from one of the current sources of Figure 3.9 to the output voltage at the frequency $\omega_0+\Delta\omega$ is then easily calculated to be [47]

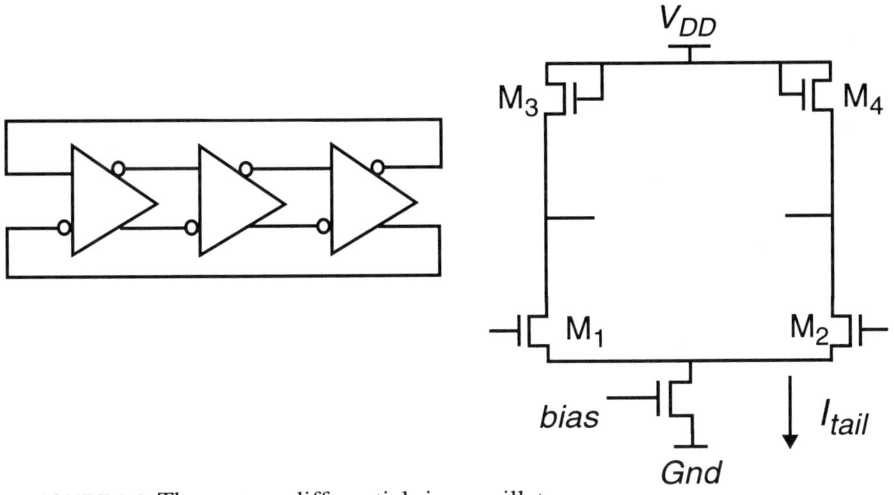

FIGURE 3.8 Three stage differential ring oscillator.

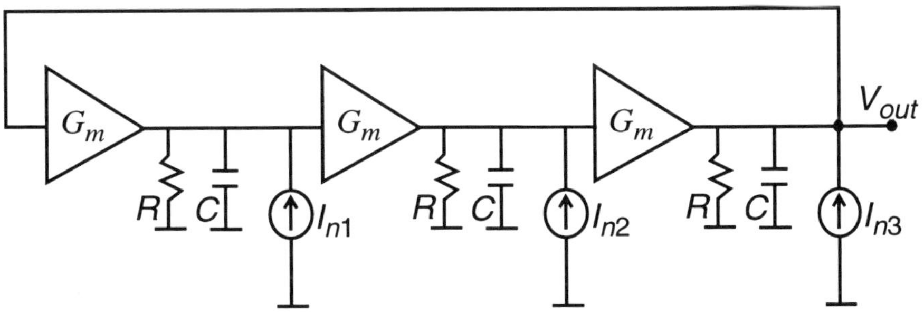

FIGURE 3.9 The small signal single-ended equivalent for the ring of Figure 3.8.

$$\left|\frac{V_{out}}{I_n}(\omega_0 + \Delta\omega)\right|^2 = \frac{R^2}{27}\left(\frac{\omega_0}{\Delta\omega}\right)^2 \tag{3.9}$$

Using the standard long channel expression for drain current noise, *i.e.*, $i_n^2/\Delta f = 8kT(g_{m1} + g_{m3})/3 \approx 8kT/R$, the output power density will be given by [47]

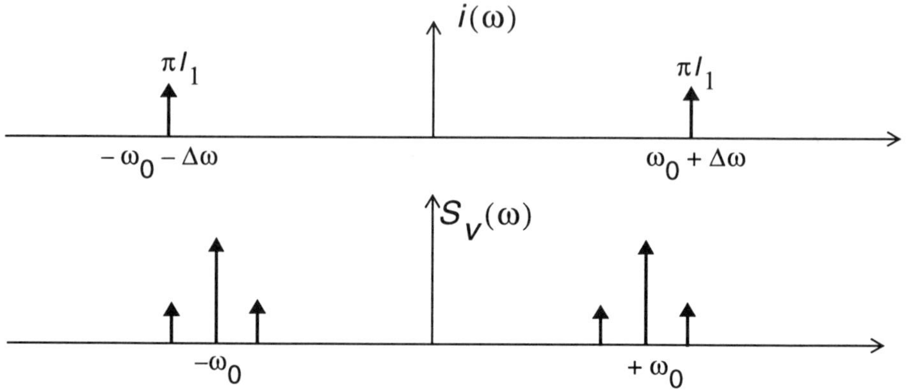

FIGURE 3.10 Pair of sidebands generated by injection $\Delta\omega$ away from the carrier.

$$\frac{\overline{V_{out}^2}}{\Delta f}(\omega_0 + \Delta\omega) = \frac{8kTR}{9}\left(\frac{\omega_0}{\Delta\omega}\right)^2 \tag{3.10}$$

There are three such noise sources in the circuit. Therefore, following the assumption in [47] that these is no correlation between the sources, the total output power density will be three times (3.10). Dividing the noise power by the carrier power, $V_{swing}^2/2$, an expression for the phase noise due to the channel noise is obtained:

$$\mathcal{L}\{\Delta\omega\} = 10 \cdot \log\left[\frac{16}{3}\frac{kTR}{V_{swing}^2}\left(\frac{\omega_0}{\Delta\omega}\right)^2\right] = 10 \cdot \log\left[\frac{8}{3}\frac{kT}{P_{load}}\left(\frac{\omega_0}{\Delta\omega}\right)^2\right] \tag{3.11}$$

where P_{load} is the power dissipated in the load device.

Although this approach results in a simple expression, it is limited to differential ring oscillators with a small number of stages because (3.9) assumes linearity of the input current-to-output voltage transfer function. This assumption only holds in differential ring oscillators with a small number of stages and small voltage swing[1]. It also assumes that the system is time invariant. As mentioned in [47], injection of a small sinusoidal current at $\omega_0+\Delta\omega$ into a real oscillator results in a *pair* of *equal* sidebands at $\omega_0 \pm \Delta\omega$, as shown in Figure 3.10. An LTI system cannot predict these sidebands

1. As will be seen in the next chapter, it is possible to introduce a different set of input and output variables that do not face these limitations.

since it is incapable of having an output at frequencies other than the input frequency or the frequency of its poles. The mechanism that results in these sidebands as well as their equality will be discussed in CHAPTER 4.

CHAPTER 4

Time-Variant Phase Noise Model

This chapter introduces the time-variant phase noise model. Section 4.1 defines the phase impulse response and uses it to evaluate the phase perturbation of an oscillator due to a small source. In Section 4.2, this approach is applied to find the output of an oscillator for the specific case of a single tone perturbation. Extension to the case of phase noise and jitter due to a random noise source is performed in Section 4.3.

Precisely how low frequency noise becomes close-in phase noise is unclear in older models, so Section 4.4 details the mechanism for upconversion of low frequency noise. Methods for properly modeling the effect of cyclostationary noise sources and time-varying elements are presented in Section 4.5. Modeling of amplitude noise, as opposed to phase noise, is investigated in Section 4.6. Finally, it is shown in Section 4.7 that the time-variant phase noise model subsumes older phase noise models.

4.1 Impulse Response Model for Excess Phase

In the general case, multiple noise sources affect the phase and amplitude of an oscillator. This chapter begins by investigating the effect of a single noise source on the amplitude and phase of the oscillator. Generalization to the case of multiple input sources with arbitrary correlations will be discussed in CHAPTER 7.

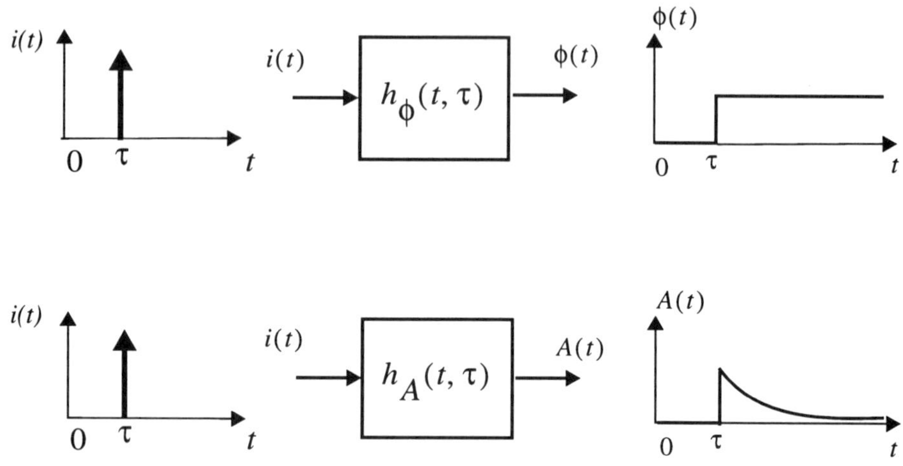

FIGURE 4.1 Equivalent systems for phase and amplitude.

Since each input source generally affects both amplitude and phase, a pair of equivalent systems, one each for amplitude and phase, can be defined. Each system can be viewed as a single-input, single-output system as shown in Figure 4.1. The input of each system in Figure 4.1 is a perturbation current (or voltage) and the outputs are the excess phase, $\phi(t)$, and amplitude, $A(t)$, as defined by (2.1). Both systems shown in Figure 4.1 are time-variant as shown by the following examples.

The first example is an ideal parallel LC tank oscillating with a voltage amplitude, V_{max}, as shown in Figure 4.2. If one injects an impulse of current at the voltage maximum, only the voltage across the capacitor changes; there is no effect on the current through the inductor. Therefore, the tank voltage changes instantaneously, as shown in Figure 4.2. Assuming a voltage- and time-invariant capacitor[1], the instantaneous voltage change ΔV is given by

$$\Delta V = \frac{\Delta q}{C_{total}} \tag{4.1}$$

where Δq is the total charge injected by the current impulse and C_{total} is the total capacitance in parallel with the current source. It can be seen from Figure 4.2 that the resultant change in $A(t)$ and $\phi(t)$ is time dependent. In particular, if the impulse is

1. The general case of time variant elements is considered in Section 4.5.2.

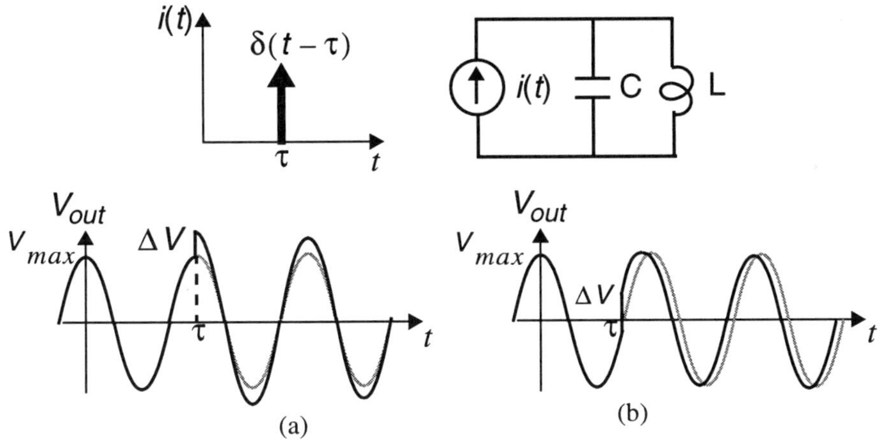

FIGURE 4.2 Impulse response of an ideal LC oscillator.

applied at the peak of the voltage across the capacitor, there will be no phase shift and only an amplitude change will result, as shown in Figure 4.2a. On the other hand, if this impulse is applied at the zero crossing, it has the maximum effect on the excess phase, $\phi(t)$, and the minimum effect on the amplitude, as depicted in Figure 4.2b.

In any practical stable oscillator, some form of amplitude restoring mechanism exists. This amplitude limiting may be due to an explicit automatic gain control (AGC) or the intrinsic nonlinearity of the devices, and results in an important difference between the phase and amplitude responses of practical oscillators. In response to a current impulse, the excess amplitude undergoes some transient behavior but finally converges to zero. However, fluctuations in the excess phase are not quenched by any restoring mechanism and therefore persist indefinitely.

Based on the foregoing argument, a current impulse results in a step change in phase, as shown in Figure 4.1. The height of this step will depend on the instant of charge injection as underlined in Figure 4.2. It is important to note that regardless of the size of the injected charge, the systems in Figure 4.1 remain time-variant.

The concept of amplitude restoration can also be visualized in the state-space portrait of the oscillator shown in Figure 4.3. The effect of this restoring mechanism is pictured as a closed trajectory in state-space. The state of the system finally approaches this trajectory, called a limit cycle, irrespective of its starting point [6]-[12].

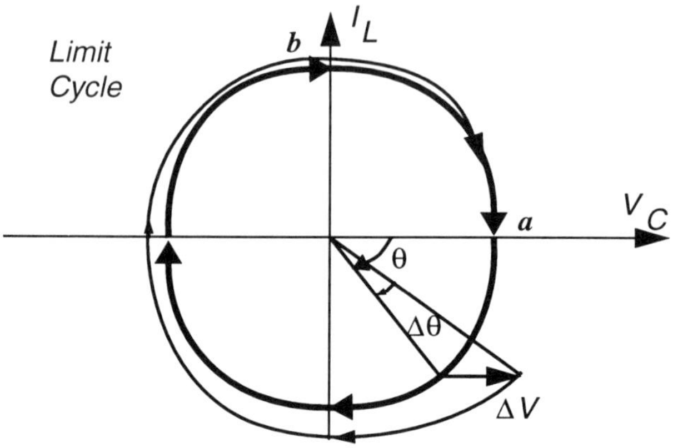

FIGURE 4.3 Limit-cycle due to amplitude restoring mechanism.

Time dependence can also be observed in the state-space trajectory shown in Figure 4.3. Applying an impulse at the peak of the capacitor voltage is equivalent to a sudden jump in voltage at point ***a***, which results in the minimum phase change and the maximum amplitude change, while applying an impulse at point ***b*** has the opposite effect. An impulse applied sometime between these two extremes will result in both amplitude and phase changes.[1]

To emphasize the generality of this time-variance, consider two more examples. The relaxation oscillator known as the Bose oscillator is shown in Figure 4.4. It consists of a Schmitt-trigger inverter and an RC circuit. The hysteresis in the transfer function of the inverter and the RC time constant determine the frequency of oscillation. The resulting capacitor voltage waveform is shown with a solid line in Figure 4.5.

As before, imagine an impulsive current source in parallel with the capacitor, injecting charge at $t=\tau$, as shown in Figure 4.4. All of the injected charge goes into the capacitor and changes the voltage across it instantaneously. This voltage change, ΔV, results in a phase shift, $\Delta\phi$, as shown in Figure 4.5.

As can be seen from Figure 4.5, for a small area of the current impulse (injected charge), the resultant phase shift is proportional to the voltage change, ΔV, and hence to the injected charge, Δq. Therefore, $\Delta\phi$ can be written as

1. Note that the angle θ in Figure 4.3 is not necessarily equal to the phase, ϕ, because the state of the system does not necessarily traverse the limit cycle with constant "angular velocity". More discussion on this distinction will appear in CHAPTER 7.

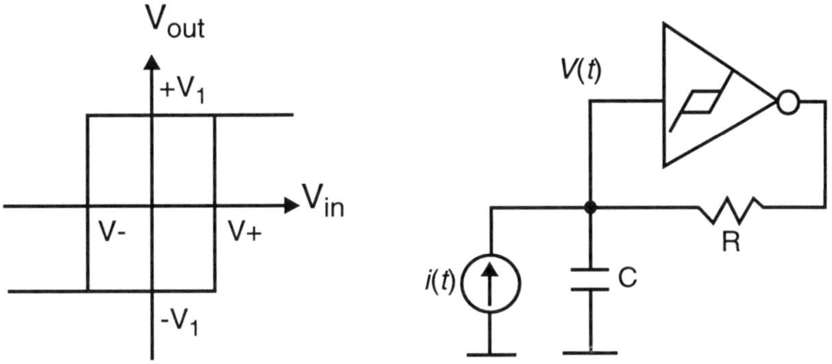

FIGURE 4.4 Bose oscillator with parallel perturbation current source.

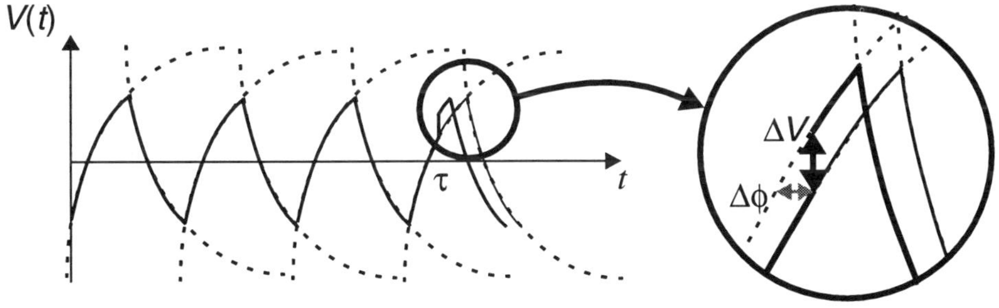

FIGURE 4.5 The waveform of the Bose oscillator shown in Figure 4.4.

$$\Delta\phi = \Gamma(\omega_0\tau)\frac{\Delta V}{V_{max}} = \Gamma(\omega_0\tau)\frac{\Delta q}{q_{max}} \qquad \Delta q \ll q_{max} \tag{4.2}$$

where V_{max} is the voltage swing across the capacitor and $q_{max} = C_{node}V_{max}$ is the maximum charge swing. The function, $\Gamma(x)$, is the time-varying "proportionality factor". It is called the *impulse sensitivity function* (ISF), since it determines the sensitivity of the oscillator to an impulsive input[1]. It is a dimensionless, frequency- and amplitude-independent function periodic in 2π that describes how much phase shift results from applying a unit impulse at any point in time.

1. The calculation of the ISF is the subject of Appendix D.

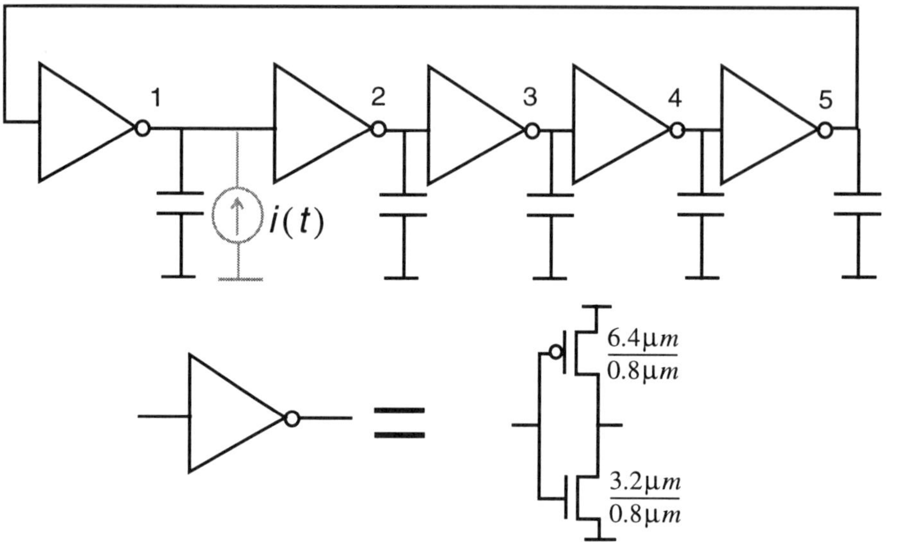

FIGURE 4.6 Five stage 1GHz ring oscillator with CMOS inverters.

The concept of the ISF can be developed further with a third example, a ring oscillator. Consider the single-ended inverter-chain ring oscillator shown Figure 4.6 with a single current source on one of its nodes. Suppose that the current is an impulse with area Δq (in coulombs), occurring at time $t=\tau$. Charge injection causes an instantaneous change in the voltage of that node, by an amount given by (4.1). This in turn produces a shift in the transition time, by an amount proportional to the injected charge (for small Δq). Again, this phase shift is time variant as can be seen from the simulated waveforms of the oscillator shown in Figure 4.7. The induced phase shift can be expressed using (4.2), with a different ISF.

To illustrate the significance of the ISF, the waveform and the ISF for the three examples are shown in Figure 4.8. As can be seen, the ISF for an ideal LC oscillator with a cosine waveform is a sine function. This result may be shown by directly solving the differential equation in Appendix C. In the case of the Bose oscillator, the ISF is inversely proportional to the slope of the capacitor voltage. For the ring oscillator, the ISF is maximum during transitions and minimum during the times that the stage is saturated to the supply or ground.

It is critical to note that the current-to-phase transfer function is linear for small injected charge, even though the active elements may have strongly nonlinear voltage-current behavior. It should also be noted that the linearity and time-variance of a sys-

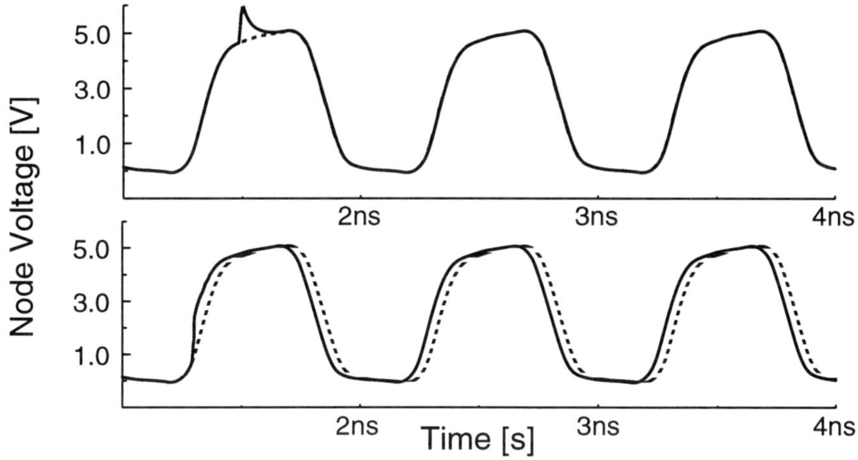

FIGURE 4.7 Simulated waveform for injection during transition and peak.

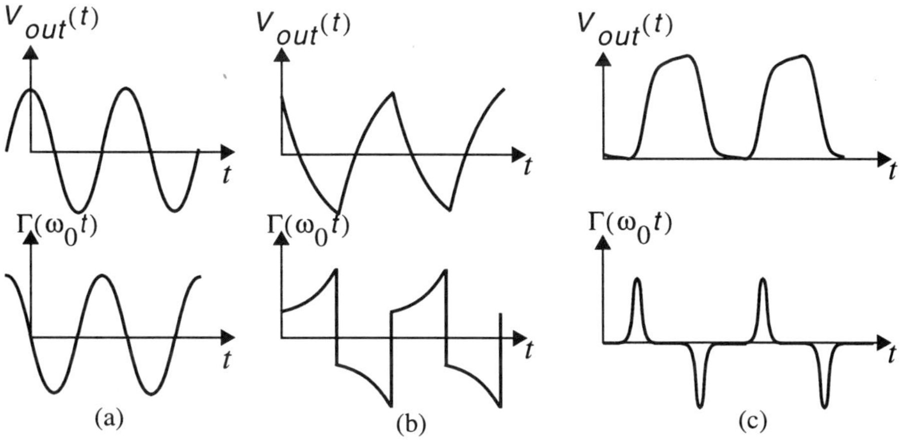

FIGURE 4.8 Typical ISF for (a) LC, (b) Bose and (c) ring oscillators.

tem depends on both the characteristics of the system and its input and output variables. The linearization of the current-to-phase system of Figure 4.1 does not imply linearization of the nonlinearity of the voltage-current characteristics of the active devices. In fact, this nonlinearity affects the shape of the ISF and therefore has an important influence on phase noise, as will be seen shortly.

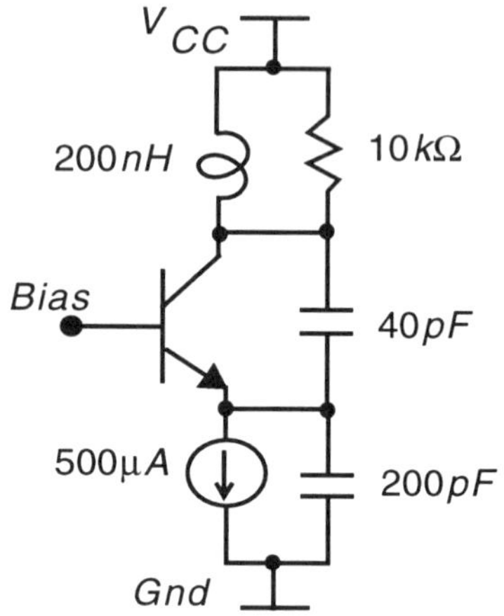

FIGURE 4.9 Bipolar 62MHz Colpitts oscillator.

The validity of the linearity assumption for small perturbations will be demonstrated more formally in CHAPTER 7. However, the extent to which this linearity assumption is valid can be investigated informally through simulation by injecting impulses with different areas (charges) and measuring the resultant phase change. SPICE simulations for the CMOS inverter-chain ring oscillator of Figure 4.6 and the Colpitts oscillator of Figure 4.9 are shown in Figure 4.10a and b, respectively. The impulse is applied close to a zero crossing, where it has the maximum effect on phase. The maximum injected charges, Δq, in Figure 4.10a and b are 25 and 16 percent of the maximum charge swings, q_{max}, respectively. Noting that the effective injected charges due to actual noise and interference sources in practical circuits are several orders of magnitude smaller than the maximum amounts of charge injected in Figure 4.10, these simulations verify the validity of linearity assumption in these oscillators. Thus, as long as the injected charge is small, the equivalent systems for amplitude and phase shown in Figure 4.1 can be fully characterized using their linear time-variant unit impulse responses, $h_\phi(t,\tau)$ and $h_A(t,\tau)$.

Noting that the introduced phase shift persists indefinitely, the unity phase impulse response can be easily calculated from (4.2) to be

$$h_\phi(t,\tau) = \frac{\Gamma(\omega_0\tau)}{q_{max}}u(t-\tau) \quad \textbf{(4.3)}$$

where $u(t)$ is the unit step.

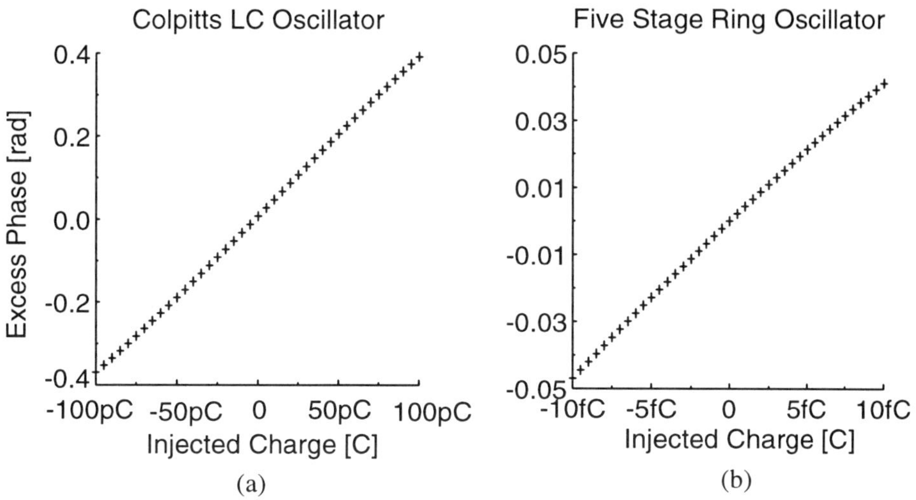

FIGURE 4.10 The simulated phase shift for (a) Colpitts oscillator and (b) ring oscillator.

Thanks to linearity, the output excess phase, $\phi(t)$, can be calculated for small charge injections using the superposition integral[1]

$$\phi(t) = \int_{-\infty}^{\infty} h_\phi(t, \tau) i(\tau) d\tau = \int_{-\infty}^{t} \frac{\Gamma(\omega_0 \tau)}{q_{max}} i(\tau) d\tau \tag{4.4}$$

where $i(t)$ represents the input noise current injected into the node of interest. Equation (4.4) is one of the most important results of this section and will be referred to frequently.

The output voltage, $V(t)$, is related to the phase, $\phi(t)$, through a phase modulation process. Thus the complete process by which a noise input becomes an output perturbation in $V(t)$ can be summarized in the block diagram of Figure 4.11. The essential features of the block diagram of Figure 4.11 are a modulation by a periodic function, an ideal integration and a nonlinear phase modulation. The complete process thus can

1. In the most general case, the argument of the ISF is shifted by the same $\phi(\tau)$ that shifts the waveform, therefore, it should be $\Gamma[\omega_0\tau+\phi(\tau)]$. However, for the most practical cases, the variations of $\phi(\tau)$ are much slower and smaller than $\omega_0\tau$ and can be ignored.

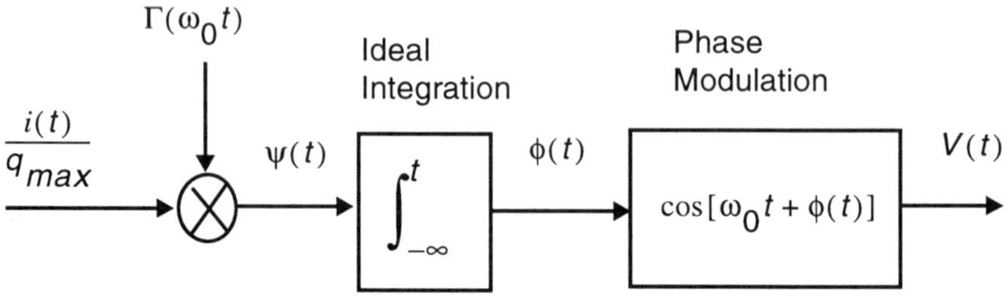

FIGURE 4.11 The equivalent block diagram of the process.

be viewed as a cascade of an LTV system that converts current (or voltage) to phase, with a nonlinear system that converts phase to voltage.

Since the ISF is periodic, it can be expanded in a Fourier series

different definition from [43] eq 12

$$\Gamma(\omega_0\tau) = c_0 + \sum_{n=1}^{\infty} c_n \cos(n\omega_0\tau + \theta_n) \tag{4.5}$$

where the coefficients c_n are real-valued, and θ_n is the phase of the nth harmonic. As will be seen later, θ_n is not important for random input noise and is thus neglected here. Using the expansion in (4.5) for $\Gamma(\omega_0\tau)$ in the superposition integral and exchanging the order of summation and integration, the following is obtained:

$$\phi(t) = \frac{1}{q_{max}}\left[c_0 \int_{-\infty}^{t} i(\tau)d\tau + \sum_{n=1}^{\infty} c_n \int_{-\infty}^{t} i(\tau)\cos(n\omega_0\tau)d\tau\right] \tag{4.6}$$

Equation (4.6) identifies individual contributions to the total $\phi(t)$ for an arbitrary input current $i(t)$ injected into any circuit node, in terms of the various Fourier coefficients of the ISF. The decomposition implicit in (4.6) can be better understood with the equivalent block diagram shown in Figure 4.12.

Each branch of the equivalent system in Figure 4.12 acts as a bandpass filter[1] and a downconverter in the vicinity of an integer multiple of the oscillation frequency. For example, the second branch weights the input by c_1, multiplies it with a tone at ω_0 and integrates the product. Hence, it passes the frequency components around ω_0 and

1. Lowpass for the first branch.

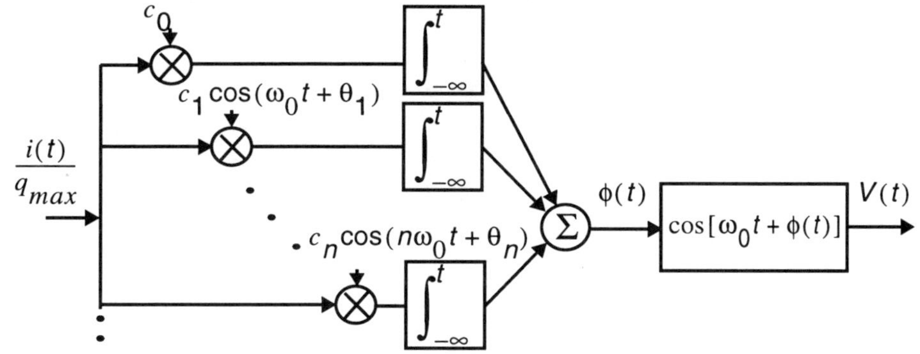

FIGURE 4.12 The equivalent system for ISF decomposition.

downconverts the output to the baseband. As can be seen, components of perturbations in the vicinity of integer multiples of the oscillation frequency play the most important role in determining $\phi(t)$.

4.2 Response to Sinusoidal Input

The response of an oscillator to small sinusoidal perturbations provides valuable information about the effect of noise at various frequencies, and is therefore investigated in this section. As mentioned previously, this process can be broken down into the two cascaded processes shown in Figure 4.11. The following subsections examine these two processes in detail.

4.2.1 Perturbation Current-to-Phase Transformation

To investigate the effect of low frequency perturbations on the oscillator phase, a low frequency sinusoidal perturbation current, $i(t)$, is injected into the oscillator at a frequency of $\Delta\omega \ll \omega_0$:

$$i(t) = I_0 \cos(\Delta\omega t) \tag{4.7}$$

where I_0 is the amplitude of $i(t)$. The arguments of all the integrals in (4.6) are at frequencies higher than $\Delta\omega$ and are significantly attenuated by the averaging nature of the integration, except the term arising from the first integral (the first branch in the

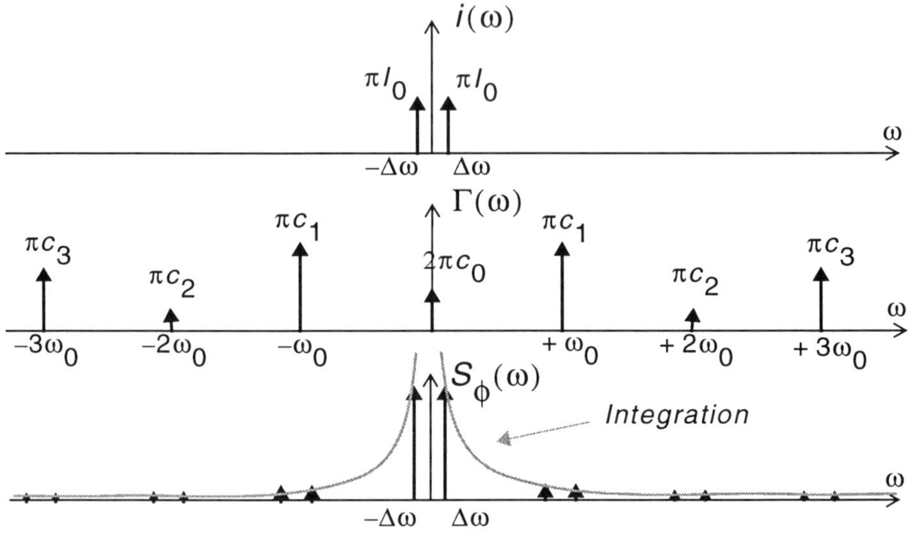

FIGURE 4.13 Conversion of a low frequency sinusoidal current to phase.

equivalent block diagram of Figure 4.12), which involves c_0. Therefore, the only significant term in $\phi(t)$ will be

$$\phi(t) \approx \frac{I_0 c_0}{q_{max}} \int_{-\infty}^{t} \cos(\Delta\omega\tau)d\tau = \frac{I_0 c_0 \sin(\Delta\omega t)}{q_{max}\Delta\omega} \quad \textbf{(4.8)}$$

As a result, there will be two impulses at $\pm\Delta\omega$ in the power spectral density of ϕ(t), denoted as $S_\phi(\omega)$ as shown in Figure 4.13[1].

As another important special case, consider a current at a frequency close to the oscillation frequency given by

$$i(t) = I_1 \cos[(\omega_0 + \Delta\omega)t] \quad \textbf{(4.9)}$$

A process similar to that of the previous case occurs except that the spectrum of $i(t)$ consists of two impulses at $\pm(\omega_0+\Delta\omega)$, as shown in Figure 4.14. This time the domi-

1. All of the impulses in Figure 4.13 and Figure 4.14 are scaled by a factor of 2π since the x-axis is the angular frequency, ω, instead of the ordinary frequency, f.

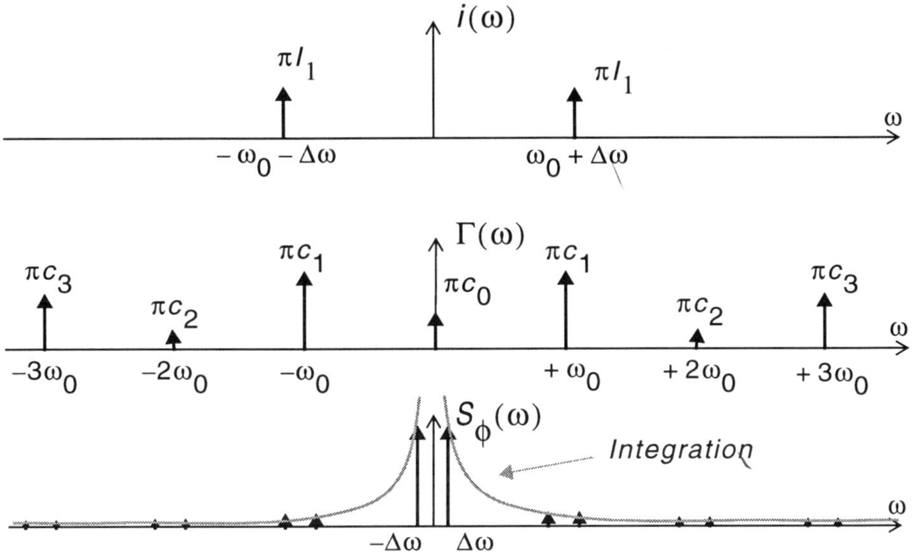

FIGURE 4.14 Conversion of a tone in the vicinity of ω_0.

nant term in (4.6) will be the second integral corresponding to n=1. Therefore, $\phi(t)$ is given by

$$\phi(t) \approx \frac{I_1 c_1 \sin(\Delta\omega t)}{2q_{max}\Delta\omega} \tag{4.10}$$

which again results in two equal sidebands at $\pm\Delta\omega$ in $S_\phi(\omega)$.

More generally, (4.6) suggests that applying a current, $i(t)=I_n\cos[(n\omega_0+\Delta\omega)t]$, close to any integer multiple of the oscillation frequency will result in two equal sidebands at $\pm\Delta\omega$ in $S_\phi(\omega)$. Hence, in the general case $\phi(t)$ is given by

$$\phi(t) \approx \frac{I_n c_n \sin(\Delta\omega t)}{2q_{max}\Delta\omega} \tag{4.11}$$

for $n \neq 0$. For $n = 0$, phase is given by (4.8).

4.2.2 Phase-to-Voltage Transformation

In the last subsection, the amount of phase error due to a given sinusoidal current was calculated to be given by (4.11). Computing the power spectral density (PSD) of the oscillator output voltage, $S_v(\omega)$, requires knowledge of how the output voltage relates to the excess phase variations.

Justify

To obtain the sideband power around the fundamental frequency, the fundamental Fourier component of the oscillator output, $\cos[\omega_0 t+\phi(t)]$, can be used as the transfer function for the last system in Figure 4.11. Note that this is a nonlinear transfer function with $\phi(t)$ as the input.

The phase-to-voltage conversion process for a single tone is now considered. For small values of $\phi(t)$, $\cos[\omega_0 t+\phi(t)]$ can be approximated as

$$\begin{aligned}\cos[\omega_0 t + \phi(t)] &= \cos(\omega_0 t)\cos[\phi(t)] - \sin(\omega_0 t)\sin[\phi(t)] \\ &\approx \cos(\omega_0 t) - \phi(t)\sin(\omega_0 t)\end{aligned} \tag{4.12}$$

where it is assumed that $\cos[\phi(t)] \approx 1$ and $\sin[\phi(t)] \approx \phi(t)$ for small values of $\phi(t)$. Using this narrowband tone modulation approximation [139], an injected current at $n\omega_0+\Delta\omega$ is seen to result in a pair of *equal* sidebands at $\omega_0 \pm \Delta\omega$. Substituting $\phi(t)$ from (4.11) into (4.12) results in a single-tone phase modulation for output voltage, whose sideband power relative to the carrier is calculated from (4.11) and (4.12) to be[1]

$$P_{dBc}(\Delta\omega) = \left(\frac{I_n c_n}{4 q_{max} \Delta\omega}\right)^2 \tag{4.13}$$

This process is shown in Figure 4.15. Appearance of the frequency deviation, $\Delta\omega$, in the denominator of the (4.13) underscores that the impulse response, $h_\phi(t,\tau)$, is a step function and therefore behaves as a time varying integrator. Equation (4.13) will be referred to frequently in subsequent sections.

The foregoing analysis predicts that a sinusoidal current injected into an oscillator at a frequency $\omega_0+\Delta\omega$ results in two equal sidebands at $\omega_0 \pm \Delta\omega$ ιν τηε ουτπυτ ϖολταγε σπεχτρυμ, as observed in [47][2]. An LTI model is not capable of predicting these sidebands, as an LTI system cannot produce any frequencies except those of the input

in the output voltage spectrum

1. For n=0, the factor of 4 in the denominator will be 8.

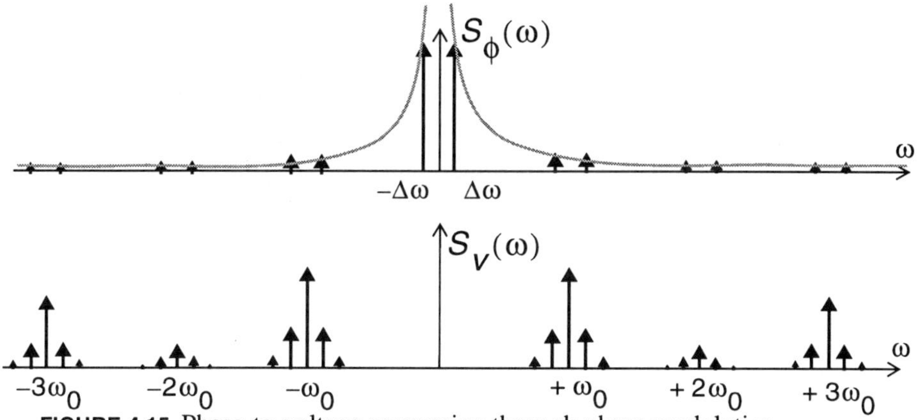

FIGURE 4.15 Phase-to-voltage conversion through phase modulation.

and those associated with the system's poles. Furthermore, the amplitude of the resulting sidebands, as well as their equality, cannot be attributed to conventional intermodulation effects. This failure is to be expected since ordinary intermodulation terms arise from nonlinearity in the voltage (or current) input/output characteristic of active devices of the form $V_{out}=\alpha_1 V_{in} + \alpha_2 V_{in}^2 + \alpha_3 V_{in}^3 +...$. This kind of effect is expected for a memoryless nonlinearity with two inputs of equal importance, but not for a system with a self-sustained mode and a small input perturbation. Furthermore, as mentioned earlier, this type of nonlinearity does not directly appear in the phase transfer characteristic and shows itself only indirectly in the shape of the ISF.

4.2.3 Simulation and Experimental Verification

It is instructive to compare the predictions of (4.13) with simulation results. A sinusoidal current of 10μA amplitude at different frequencies is injected into node 1 of the 1.01GHz ring oscillator of Figure 4.6 in HSPICE. Figure 4.16a shows the resulting simulated power spectrum of the signal on node 4 for a low frequency input at f_m=50MHz. This power spectrum is obtained using the FFT analysis capability of HSPICE 96.1 [141]. Note that the injected noise is upconverted into two equal sidebands at f_0+f_m and f_0-f_m, as predicted by (4.13). Figure 4.16b shows the effect of

2. Note that the effect of amplitude response is totally ignored here. For oscillators with large amplitude response the sidebands may show a slight difference in amplitude. Modeling of the amplitude response is described in Section 4.6.

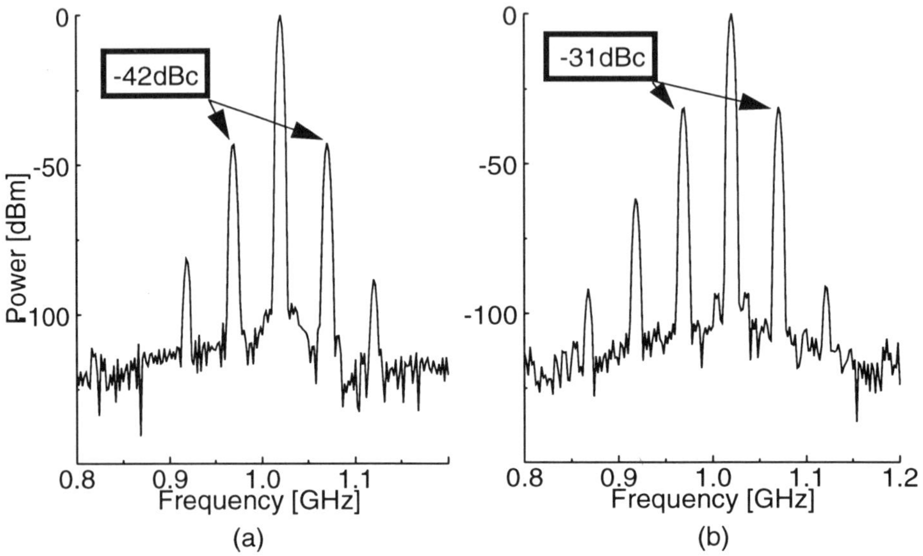

FIGURE 4.16 Simulated power spectrum of the output with current injection at (a) f_m=50MHz and (b) f_0+f_m=1.06GHz.

injection of a current at f_0+f_m=1.06GHz. Again, two equal sidebands are observed at f_0+f_m and f_0-f_m, also as predicted by (4.13).

Simulated sideband power for the general case of current injection at nf_0+f_m can be compared to the predictions of (4.13). The ISF for this oscillator is obtained by direct simulation, in which very short pulses of current are injected and the induced phase shift is measured, as described in Appendix D. Here, q_{max} is equal to $C_{node}V_{max}$, where C_{node} is approximated by the average capacitance on each node of the circuit and V_{max} is the maximum swing across it. For this oscillator, C_{node}=26fF and V_{swing}=5V, which results in q_{max}=130fC. Figure 4.17 depicts the simulated and predicted sideband powers for a sinusoidal injected current of amplitude I_n=10µA, and an f_m of 50MHz. As can be seen from the figure, these agree to within 1dB for the higher power sidebands. The discrepancy in the case of the low power sidebands (n = 4, 6-9) arises from numerical noise in the simulations, which represents a greater fractional error at lower sideband power. Overall, there is satisfactory agreement between simulation and the predictions made by (4.13).

To verify further that the conversion of perturbations from around integer multiples of the frequency of oscillation occurs as predicted by (4.13), a series of experiments on a

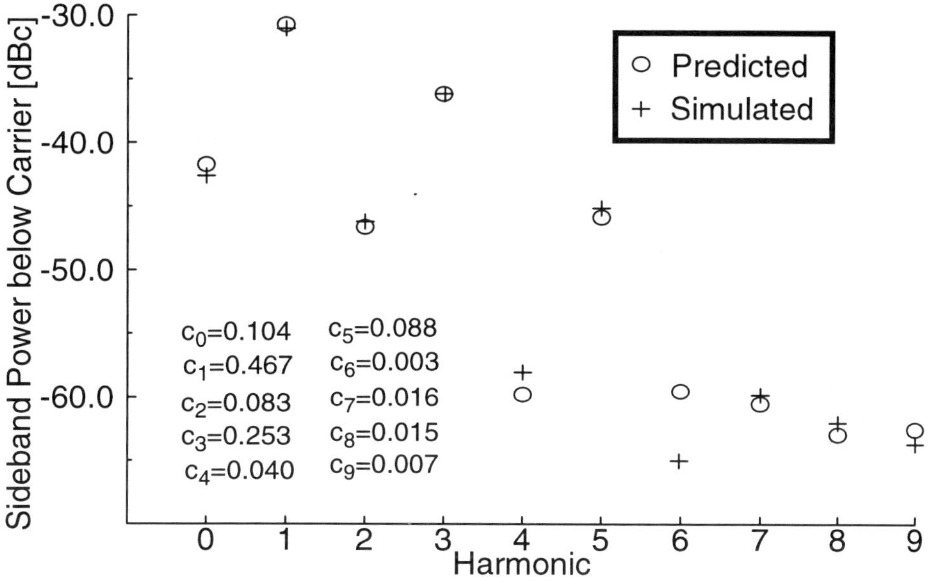

FIGURE 4.17 Simulated and calculated sideband power for the 5-stage ring of Figure 4.6.

5-stage, 5.4MHz ring oscillator constructed with ordinary CMOS inverters is performed.

The first experiment varies the frequency offset from an integer multiple of the oscillation frequency. An input sinusoidal current source of 20μA (rms) at f_m is applied to one node and the output is measured at another node. This process is repeated for f_0+f_m, $2f_0+f_m$ and $3f_0+f_m$, as shown graphically in Figure 4.18. The sideband power is measured using a spectrum analyzer and plotted versus f_m in Figure 4.19. Note that the slope in all four cases is -20dB/decade, again in complete accordance with (4.13).

The second experiment examines the linearity of the current-to-phase conversion. A sinusoidal current is injected at frequencies f_m=100kHz, f_0+f_m=5.5MHz, $2f_0+f_m$=10.9MHz and $3f_0+f_m$=16.3MHz, and the sideband powers at $f_0 \pm f_m$ are measured as the magnitude of the injected current is varied. At any amplitude of injected current, the two sidebands are equal in amplitude to within the accuracy of the spectrum analyzer (0.3dB), in complete accordance with the theory. These sideband powers are plotted versus the injected input current in Figure 4.20. As can be seen the power transfer function for the input current to the output sideband is linear as suggested by (4.13). The slope of the best fit line is 19.8dB/decade. The predicted slope

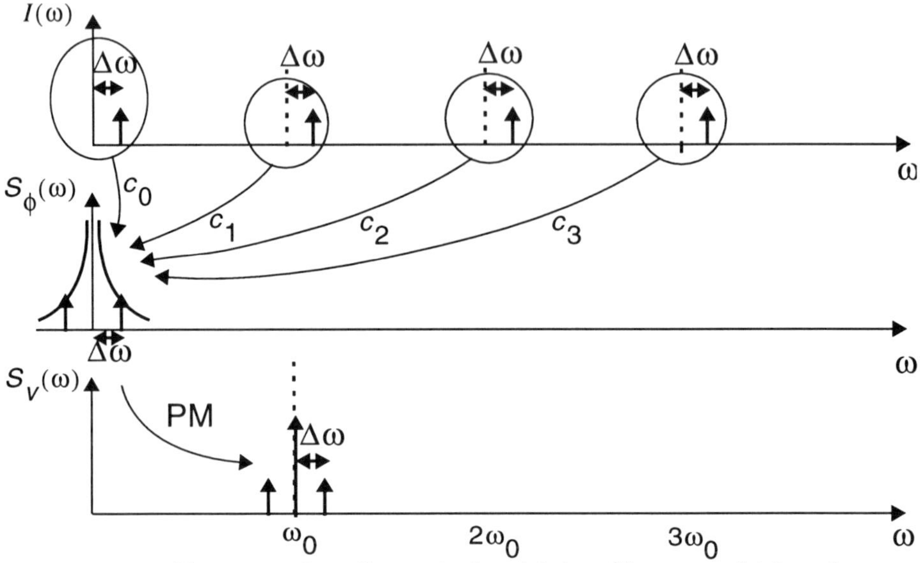

FIGURE 4.18 The conversion of tones in the vicinity of integer multiples of ω_0.

is 20dB/decade because the excess phase, $\phi(t)$, is proportional to $i(t)$ and hence the sideband power is proportional to I^2, leading to a 20dB/decade slope. The behavior shown in Figure 4.20 verifies that the underlying linearity assumption leading to (4.13) holds for injected input currents orders of magnitude larger than typical noise currents.

The third experiment verifies the effect of maximum charge swing, q_{max}, on the sideband powers. An input sinusoidal current source of 20μA (rms) at f_0+f_m is applied to one node and the output is measured at another node. The maximum charge swing, q_{max}, is doubled by adding another set of identical inverters in parallel with the first set. The experiment is repeated and, as can be seen in Figure 4.21, there is a 6dB reduction in the sideband power, in complete accordance with (4.13).

4.3 Phase Noise and Jitter due to Random Noise

In this section, the extension of the theory to the case of a *random* noise source is considered from an intuitive standpoint (a more rigorous treatment is presented in Appen-

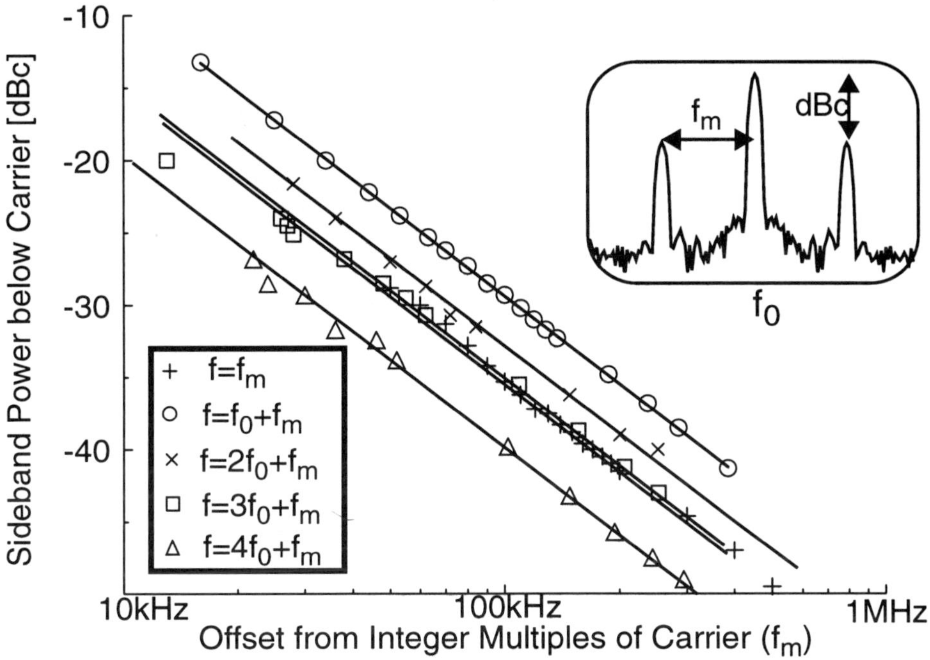

FIGURE 4.19 Measured sideband power vs. f_m for injection in vicinity of nf_0.

dix B). The model is expanded to the more general case of multiple correlated cyclostationary sources in CHAPTER 7.

4.3.1 Phase Noise

Consider a random noise current source, $i_n(t)$, whose power spectral density has both a flat region and a $1/f$ region, as shown in Figure 4.22. Equation (4.6) shows that noise components located near integer multiples of the oscillation frequency are weighted by Fourier coefficients of the ISF and integrated to form the low frequency noise sidebands for $S_\phi(\omega)$. These sidebands in turn become close-in phase noise in the spectrum of $S_v(\omega)$ through phase modulation (PM), as illustrated in Figure 4.22.

It can be seen that the total $S_\phi(\omega)$ is given by the sum of phase noise contributions from device noise in the vicinity of the integer multiples of ω_0, weighted by the coefficients c_n. This is shown in Figure 4.23 which shows the spectrum of $\phi(t)$ on *log-log* scales.

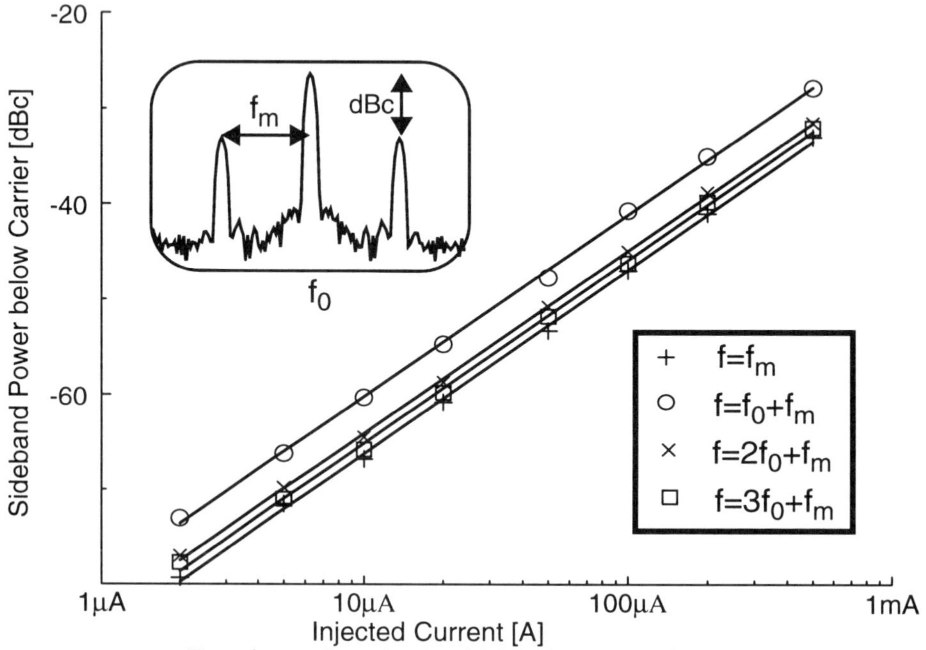

FIGURE 4.20 Experimental results for sideband power *vs.* injected current at f_m=100kHz, f_0+f_m=5.5MHz, $2f_0+f_m$=10.9MHz, $3f_0+f_m$=16.3MHz.

The theory predicts the existence of $1/f^3$ and $1/f^2$ regions in the phase noise power spectrum shown in Figure 4.23. Low-frequency noise, such as flicker noise, is weighted by the coefficient c_0 and ultimately produces a $1/f^3$ phase noise region. White noise terms are weighted by other c_n coefficients and give rise to the $1/f^2$ region of phase noise spectrum[1]. The total sideband noise power is the sum of the individual terms, as shown by the bold line in the same figure.

Retaining the narrowband PM assumption used earlier, we expect the single-sideband spectral noise density, $\mathcal{L}\{\Delta\omega\}$, to be similar to the spectrum $\phi(t)$, *i.e.*, $S_\phi(\omega)^2$. To verify this expectation, we now carry out a quantitative analysis of the phase noise sideband power. Consider a white input noise current with power spectral density $\overline{i_n^2}/\Delta f$. Note that I_n in (4.13) represents the peak and not the rms amplitude, hence, $I_n^2/2 = \overline{i_n^2}/\Delta f$

1. It is apparent that if the original noise current, $i(t)$, contains $1/f^n$ low frequency noise terms, they will appear in the phase noise spectrum as $1/f^{n+2}$ regions.

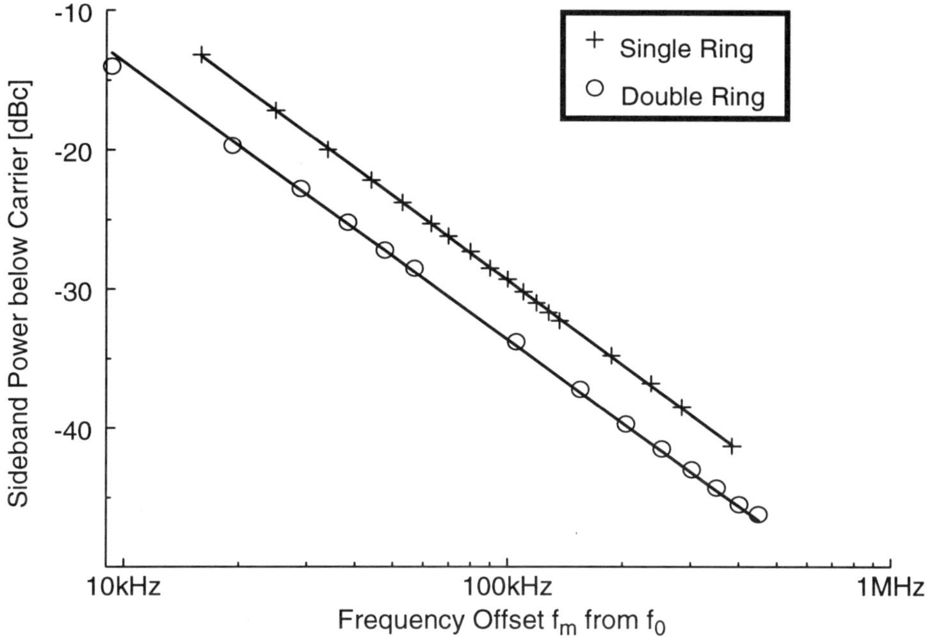

FIGURE 4.21 Sideband power for single and double ring oscillators.

for Δf=1Hz. Noise power around the frequency $n\omega_0 + \Delta\omega$ causes two equal sidebands at $\omega_0 \pm \Delta\omega$, as shown in Figure 4.14. However, it is important to recognize that noise power at $n\omega_0 - \Delta\omega$ also has a similar effect. Therefore, twice the power of noise at $n\omega_0 + \Delta\omega$ should be taken into account. Based on the foregoing development and (4.13), the total single-sideband phase noise spectral density due to one noise source at an offset frequency $\Delta\omega$ is given by the sum of the powers of the highlighted components in Figure 4.22, namely

Error in [43] eq 19 not mentioned

2. A more accurate analysis that takes into account the effect of wideband PM is given in Appendix B. That analysis shows that, although the phase power spectrum, $S_\phi(\omega)$, grows without bound as the frequency goes to zero, the power spectrum of the output voltage, $S_v(\omega)$, usually reaches a plateau. This result is intuitively satisfying, since $\phi(t)$ becomes $V(t)$ through the cosine function, which limits its span and, hence, the low frequency power.

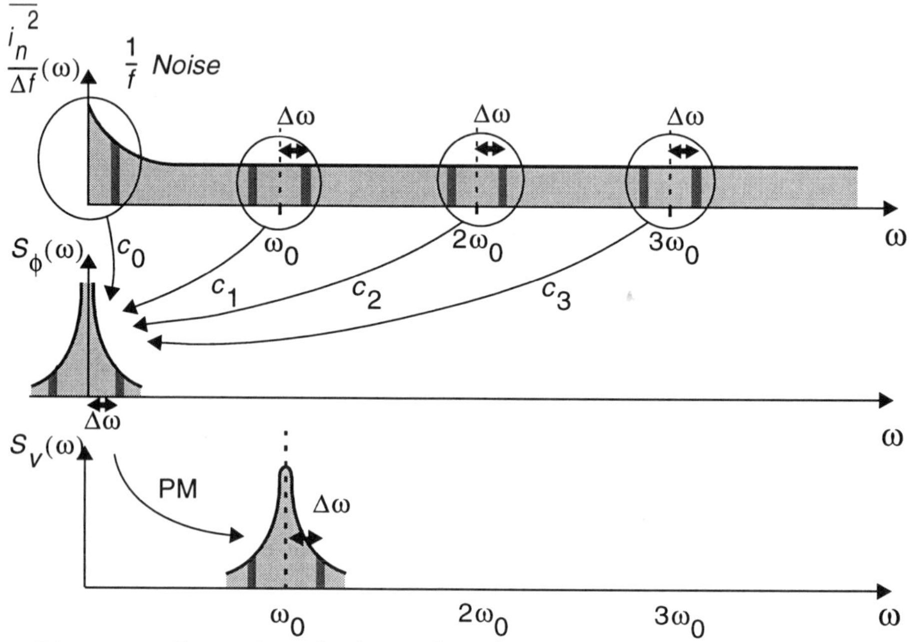

FIGURE 4.22 Conversion of noise to phase fluctuations and phase-noise sidebands.

$$\mathcal{L}\{\Delta\omega\} = 10\log\left(\frac{\frac{\overline{i_n^2}}{\Delta f}\sum_{n=0}^{\infty} c_n^2}{4q_{max}^2\Delta\omega^2}\right) \tag{4.14}$$

Now, according to Parseval's relation,

$$\sum_{n=0}^{\infty} c_n^2 = \frac{1}{\pi}\int_0^{2\pi}|\Gamma(x)|^2 dx = 2\Gamma_{rms}^2 \tag{4.15}$$

where Γ_{rms} is the rms value of $\Gamma(x)$. As a result

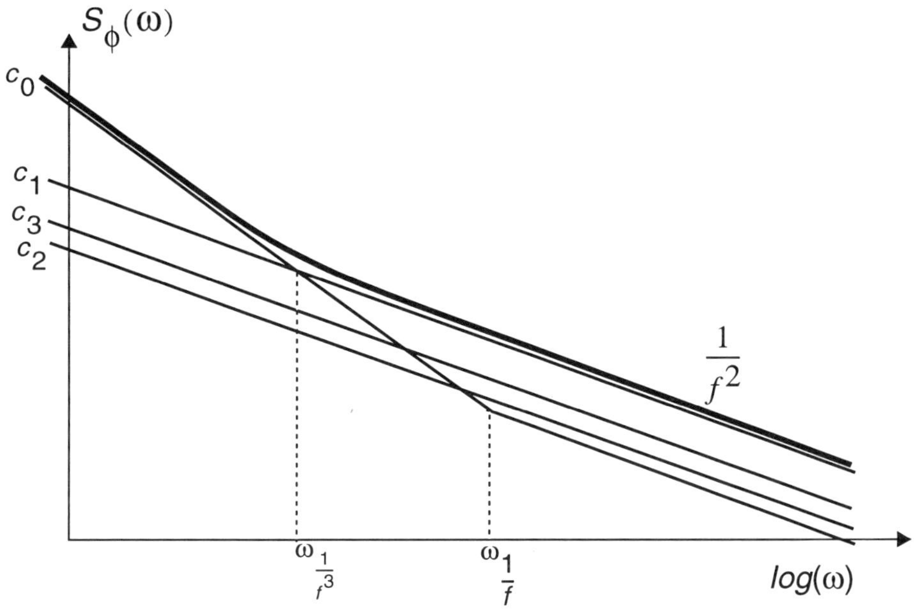

FIGURE 4.23 $S_\phi(\omega)$ on a log-log axis.

$$\mathcal{L}\{\Delta\omega\} = 10\log\left(\frac{\Gamma_{rms}^2}{q_{max}^2} \cdot \frac{\overline{i_n^2}/\Delta f}{2 \cdot \Delta\omega^2}\right) \tag{4.16}$$

This equation gives the phase noise spectrum of an arbitrary oscillator in the $1/f^2$ region of the phase noise spectrum and will be referred to frequently. Note the absence of any fitting parameters.

So far the case of a perturbation current source in parallel with a capacitor has been analyzed. The dual case of a voltage source in series with an inductor can also occur in an oscillator. The series tank shown in Figure 4.24 can be used to calculate the effect of a voltage noise source. An impulse of voltage can only change the current of the inductor. Therefore, for a noise voltage source in series with an inductor, q_{max} should be replaced with $p_{max}=LI_{max}$, where p_{max} represents the maximum magnetic flux swing in the inductor, L, and I_{max} is the maximum current swing in the inductor. Therefore, the phase noise due to such source is given by

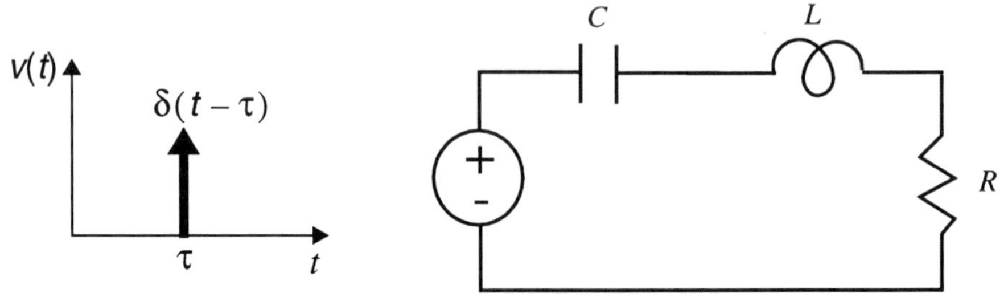

FIGURE 4.24 A voltage impulse in series with an inductor.

$$\mathcal{L}\{\Delta\omega\} = 10\log\left(\frac{\Gamma_{rms}^2}{p_{max}^2} \cdot \frac{\overline{v_n^2}/\Delta f}{2 \cdot \Delta\omega^2}\right) \tag{4.17}$$

where $\overline{v_n^2}/\Delta f$ is the voltage mean-square density per unit bandwidth and Γ_{rms} is the rms value of the ISF for the voltage source.

4.3.2 Phase Jitter

As mentioned in CHAPTER 2, timing jitter is the preferred parameter for quantifying frequency instabilities in digital circuits. The superposition integral (4.4) can also be used to calculate the phase jitter defined by (2.8). The phase jitter can be expressed as

$$\sigma_{\Delta\phi}^2 = E\{\Delta\phi_\tau^2\} = E\{[\phi(t+\tau) - \phi(t)]^2\} \tag{4.18}$$

where $E[\ ]$ represents expected value. According to (4.4),

$$\Delta\phi_\tau = \int_0^\tau \frac{\Gamma(\omega_0 t_1)}{q_{max}} i(t_1) dt_1 \tag{4.19}$$

Therefore combining (4.18) and (4.19), phase jitter can be written as

$$\sigma_{\Delta\phi}^2 = \frac{1}{q_{max}^2} \int_0^\tau \int_0^\tau \Gamma(\omega_0 t_1)\Gamma(\omega_0 t_2) E[i(t_1) i(t_2)] dt_1 dt_2 \tag{4.20}$$

Only for white noise sources

For a white noise current source, the autocorrelation function is $E[i(t_1)i(t_2)] = \frac{1}{2}\overline{i_n^2}/\Delta f \cdot \delta(t_1 - t_2)$, therefore

$$\sigma_{\Delta\phi}^2 = \frac{1}{2}\frac{\overline{i_n^2}/\Delta f}{q_{max}^2}\int_0^{\tau}\Gamma^2(\omega_0 t_1)dt_1 \tag{4.21}$$

which is

not valid for τ < T

$$\sigma_{\Delta\phi}^2 = \frac{1}{2}\frac{\overline{i_n^2}/\Delta f}{q_{max}^2}\Gamma_{rms}^2|\tau| \qquad for \qquad \tau \gg T \;\; or \;\; \tau = mT \tag{4.22}$$

where *n* is an integer. Using (4.22) and the definition of κ in (2.6), it is calculated to be

$$\kappa = \frac{\Gamma_{rms}}{q_{max}\omega_0}\sqrt{\frac{1}{2}\frac{\overline{i_n^2}}{\Delta f}} \tag{4.23}$$

This result will be used in CHAPTER 5 to calculate the timing jitter in ring oscillators.

4.4 Upconversion of Low Frequency Noise

This section is dedicated to the study of the upconversion of low frequency noise. Section 4.4.1 provides an explicit expression for the $1/f^3$ noise corner in terms of the device $1/f$ noise corner and the dc and rms values of the ISF. Section 4.4.2 shows simulations and experimental verification of the results.

4.4.1 Calculation of the $1/f^3$ Noise Corner

Many active and passive devices exhibit low frequency noise with a power spectrum that is approximately inversely proportional to the frequency. It is for this reason that noise sources with this behavior are referred to as $1/f$ noise[1] [124]-[133]. In this sub-

1. It is also referred to as flicker noise

section, the relationship between the device $1/f$ corner and the phase noise $1/f^3$ corner will be investigated quantitatively. It is important to note that nothing in the foregoing development implies that the $1/f^3$ corner of the phase noise and the $1/f$ corner of the device noise are generally coincident, as is commonly assumed. In fact, from Figure 4.23, it should be apparent that the relationship between these two corner frequencies depends on the specific values of the various coefficients c_n.

Noting that device noise in the $1/f$ region can be described by

$$\overline{i_{n,1/f}^2} = \overline{i_n^2} \cdot \frac{\omega_{1/f}}{\Delta\omega} \qquad (\Delta\omega < \omega_{1/f}) \tag{4.24}$$

where $\omega_{1/f}$ is the corner frequency of device $1/f$ noise[1], (4.13) and (4.24) result in the following expression for phase noise in the $1/f^3$ portion of the phase noise spectrum:

$$\mathcal{L}\{\Delta\omega\} = 10\log\left(\frac{c_0^2}{q_{max}^2} \cdot \frac{\overline{i_n^2}/\Delta f}{2 \cdot \Delta\omega^2} \cdot \frac{\omega_{1/f}}{\Delta\omega}\right) \tag{4.25}$$

The phase noise $1/f^3$ corner, $\Delta\omega_{1/f^3}$, is the frequency where the sideband power due to the white noise given by (4.16) is equal to the sideband power arising from the $1/f$ noise given by (4.25), as shown in Figure 4.23. Solving for $\Delta\omega_{1/f^3}$ results in the following expression for the $1/f^3$ corner in the phase noise spectrum:

$$\Delta\omega_{1/f^3} = \omega_{1/f} \cdot \left(\frac{c_0}{\Gamma_{rms}}\right)^2 \tag{4.26}$$

As can be seen, the $1/f^3$ phase noise corner is not equal to the $1/f$ device noise corner, but is *smaller* by a factor equal to c_0^2/Γ_{rms}^2, where c_0 is the dc value of ISF,

$$c_0 = \frac{1}{2\pi}\int_0^{2\pi} \Gamma(x)dx \tag{4.27}$$

1. It should be noted that using MOS transistors in a switching mode can reduce their $1/f$ noise and corner frequency [131]. If the devices in the oscillator of interest undergo switching, the new $1/f$ noise corner should be used in (4.24) [132][133].

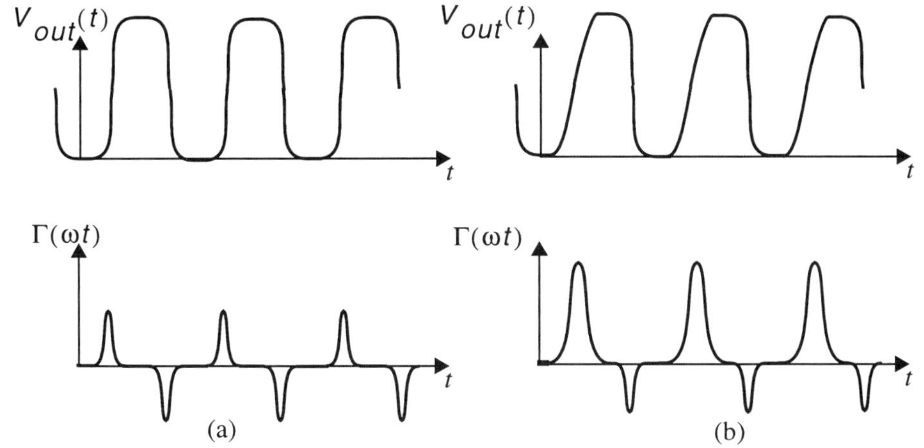

FIGURE 4.25 (a) Waveform and ISF for (a) the symmetric waveform (b) the asymmetric waveform.

To understand what affects c_0, consider two ring oscillators, with waveforms shown in Figure 4.25. The first waveform has symmetric rising and falling edges, *i.e.*, its rise-time is the same as its fall-time. Assuming a time-invariant node capacitor[1], the sensitivity of this oscillator to a perturbation during the rising edge is the same as its sensitivity during the falling edge, except for a sign. Therefore, the ISF has a small dc value. The second case corresponds to an asymmetric waveform with slow rising edge and a fast falling edge. In this case, the phase is more sensitive during the rising edge, and is also is sensitive for a longer time; therefore, the positive lobe of the ISF will be taller and wider as opposed to its negative lobe which is short and thinner, as shown in Figure 4.25.

The dc value of the ISF for the asymmetric rising and falling edge is much larger than that in the symmetric case, and hence a low frequency noise source injecting into it shows a stronger upconversion of low frequency noise. A limited case of the effect of odd-symmetric waveforms on phase noise has been observed in [28]. However minimizing (4.27) is more a general criterion because although odd-symmetric waveforms may have small c_0 coefficients, the class of waveforms with small c_0 is not limited to those with odd symmetry.

1. The effect of time-variant capacitors will be investigated in Section 4.5.

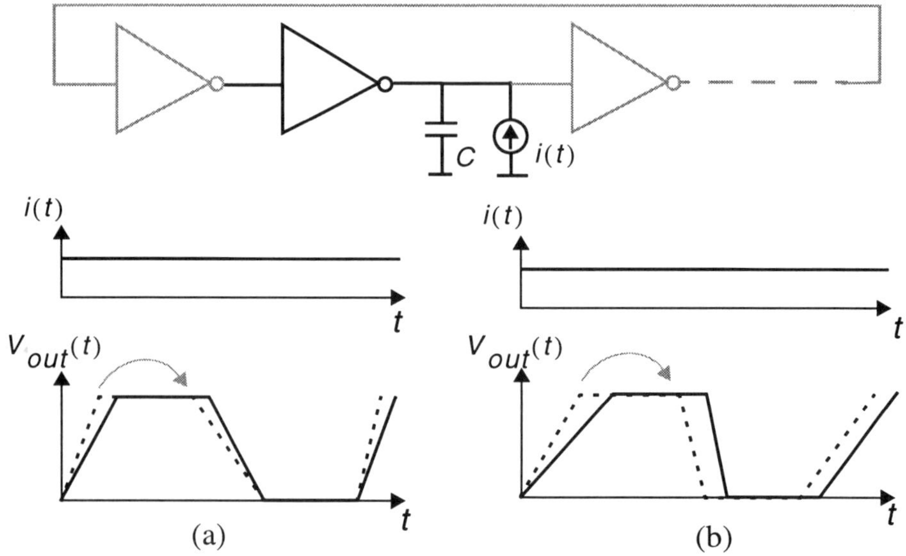

FIGURE 4.26 Effect of asymmetry in rise and fall time.

A more intuitive view of low frequency noise upconversion is shown in Figure 4.26. Consider a low frequency current source on one node of a ring oscillator, and for simplicity assume trapezoidal waveforms, as shown in Figure 4.26. The low frequency source has very small variations during one period and therefore can be approximated as a dc current, as shown in Figure 4.26.

Assuming a time-invariant capacitor, C, the rising transition will be shifted to the left by an amount determined by C, $i(t)$ and the slope of the transition. Since the falling edge is caused by the same transition propagating through the ring, it will occur earlier by the same amount, as shown by dashed lines in Figure 4.26. In the case of symmetric rising and falling edges, an equal but opposite amount of phase shift will be induced during the falling transition. Therefore, the next transition will start at the same time as before and the period is not changed. As a result, the phase variations due to low frequency noise remain local and are not integrated.

For an oscillator with asymmetric rising and falling edges, however, the phase shift introduced during the rising edge is different from the opposite phase shift introduced during the falling edge, as can be seen in Figure 4.26b. This will cause the next transition to occur earlier (or later for a negative low frequency current). This is equivalent

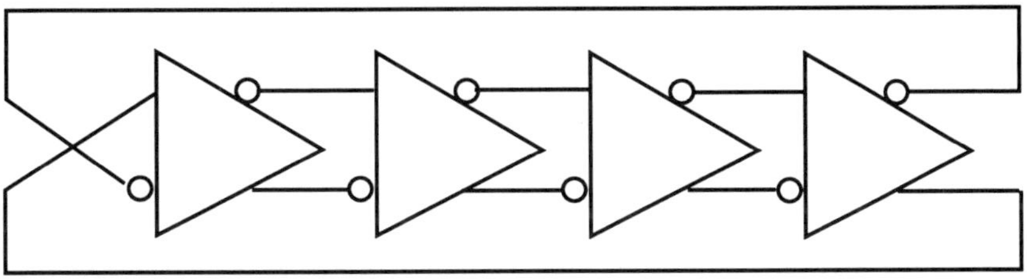

FIGURE 4.27 Four-stage differential ring oscillator.

to a net frequency change and will result in frequency (and hence phase) modulation of the high frequency signal as the low frequency source slowly changes.

Symmetry is therefore important. Recognizing this fact allows the designer to identify the design parameters that minimize the upconversion of low frequency noise, through proper device sizing, for example. The design goal should thus be the minimization of the c_0 coefficient.

The importance of symmetry might lead to the conclusion that differential signaling would minimize c_0. Unfortunately, while differential circuits are certainly symmetrical with respect to the desired signals, the differential symmetry *disappears* for the individual noise sources because they are independent of each other. Hence, it is the symmetry of each *half*-circuit that is important, as is demonstrated in the differential ring oscillator of Figure 4.27 with buffer stages shown in Figure 4.28. A sinusoidal current of 100μA at 50MHz is injected at the drain node of one of the buffer stages to model the effect of low frequency noise in the differential NMOS and PMOS devices. This sinusoidal current results in two equal sidebands, -46dB below carrier, in the power spectrum of the differential output as shown in Figure 4.29. Because of the voltage dependent conductance of the load devices, the individual waveform on each output node is not fully symmetrical and, consequently, there will be a large upconversion of noise to close-in phase noise, even though differential signaling is used. Another source whose low frequency noise is not automatically suppressed is the tail current source. This source can be the dominant source of $1/f$ noise and its effect will be discussed in detail in Section 6.3.

In the differential buffer stage of Figure 4.28, the asymmetry is due to the voltage dependent conductance of the load. Therefore, reduction of the upconversion might be achieved through the use of a more linear load, such as resistors or linearized MOS devices. These more linear loads improve the single-ended symmetry because the rising and falling behavior is governed by an RC time constant and makes the individual

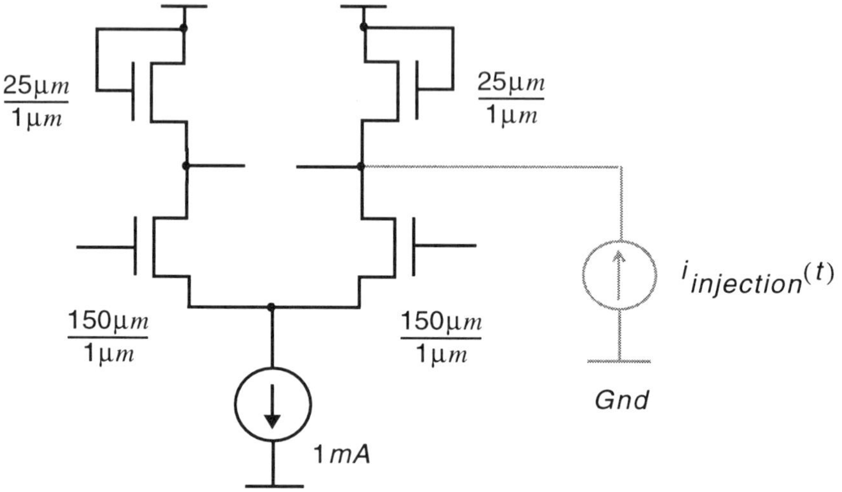

FIGURE 4.28 Buffer stage used in the four stage ring oscillator of Figure 4.27.

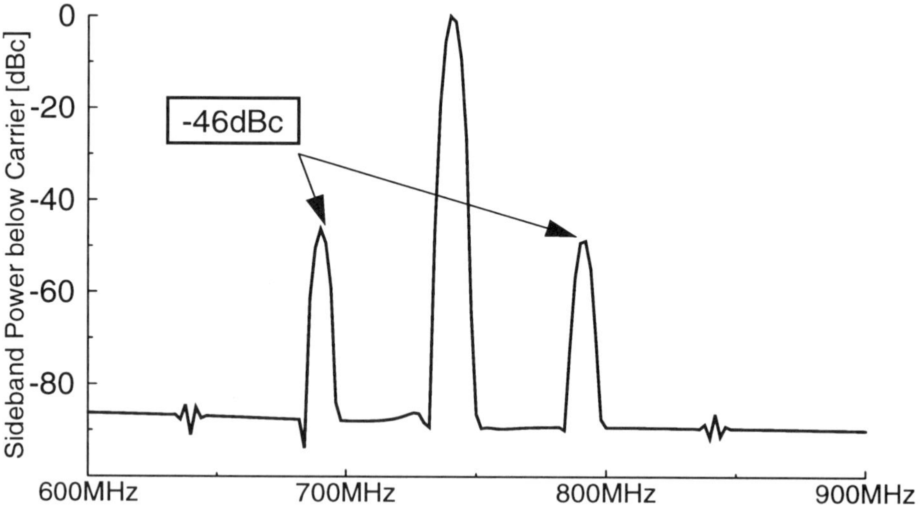

FIGURE 4.29 The induced sidebands due to current injection into a differential node.

waveforms more symmetrical. It was first observed in the context of supply noise rejection [44] that using more linear loads can reduce the effect of supply noise on

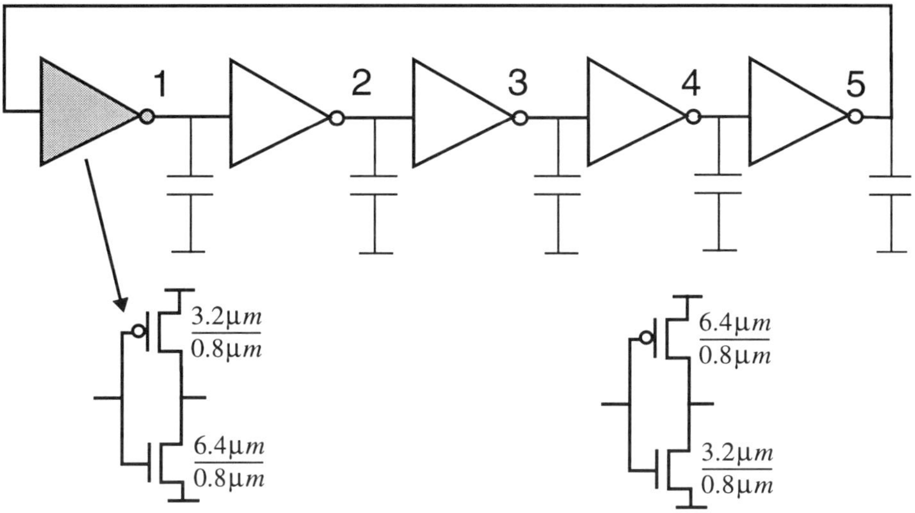

FIGURE 4.30 Ring oscillator with one asymmetric stage.

timing jitter. This treatment shows that it also improves low-frequency noise upconversion into phase noise.

4.4.2 Simulation and Experimental Verification

To illustrate the effect of a rise and fall time asymmetry, consider a purposeful imbalance of pull-up and pull-down rates in one of the inverters in the ring oscillator of Figure 4.30. This is obtained by halving the channel width, W_n, of the NMOS device and doubling the width, W_p, of the PMOS device of one inverter in the ring, as shown in Figure 4.30. The output waveform and corresponding ISFs for node 4 (a symmetric node) and node 1 (the asymmetric node), are shown in Figure 4.31.

The ISF has a large dc value for the asymmetric node compared to one of the symmetric nodes elsewhere in the ring, as depicted in Figure 4.31. From this difference in the dc values of the ISFs, it can be inferred that the close-in phase noise due to low-frequency noise sources should be smaller for the symmetrical output than for the asymmetrical one. To investigate this assertion, the results of two SPICE simulations are shown in Figure 4.32. In the first simulation, a sinusoidal current source of amplitude 10μA at f_m=50MHz is applied to one of the symmetric nodes of the oscillator. In the second experiment the same source is applied to the asymmetric node. As can be seen

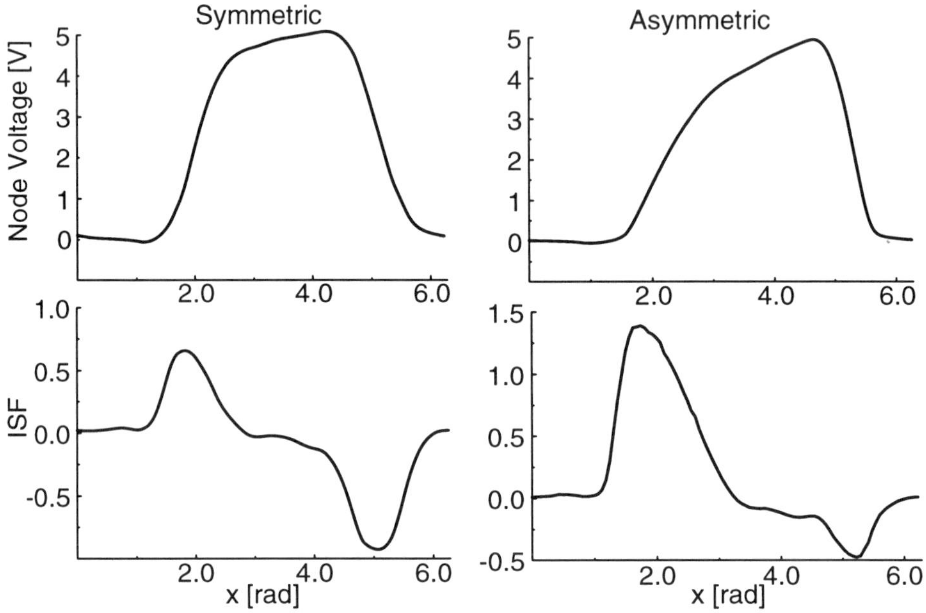

FIGURE 4.31 The waveform and the ISF for symmetric and asymmetric nodes.

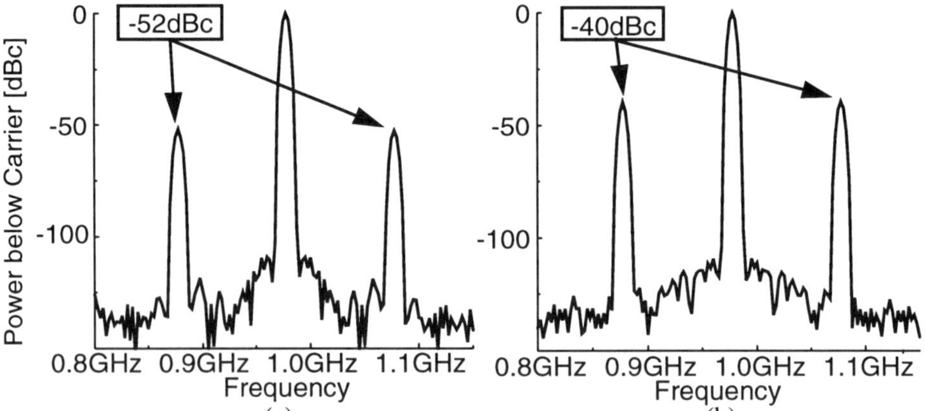

(a)
(b)

FIGURE 4.32 Simulated power spectrum with current injection at fm=50MHz into (a) symmetrical node and (b) asymmetrical node.

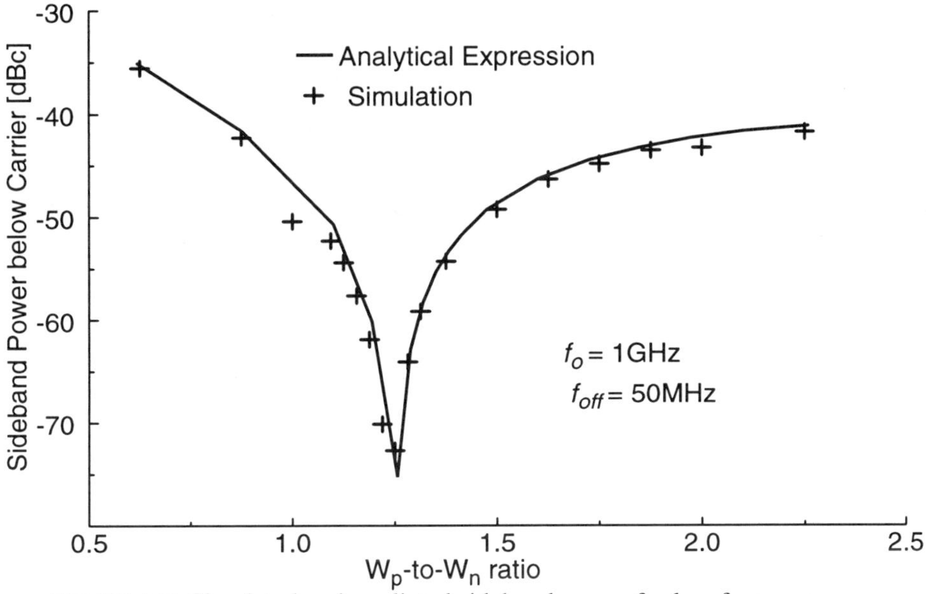

FIGURE 4.33 Simulated and predicted sideband power for low frequency injection vs. PMOS-to-NMOS *W/L* ratio of Figure 4.6.

from the power spectra, noise injected into the asymmetric node results in sidebands that are 12dB larger than at the symmetric node.

Note that (4.27) suggests that upconversion of low frequency noise can be significantly reduced by minimizing c_0. Since c_0 depends on the waveform, this observation implies that a proper choice of waveform may yield significant improvements in close-in phase noise. The following simulation explores this concept by changing the ratio of W_p-to-W_n over some range, while injecting 10μA of sinusoidal current at 100MHz into one node. The sideband power below carrier as a function of the W_p to W_n ratio is shown in Figure 4.33. The SPICE-simulated sideband power is shown with plus symbols and the sideband power as predicted by (4.13) is shown by the solid line. As can be seen, close-in phase noise due to upconversion of low-frequency noise can be suppressed by an arbitrary factor, at least in principle.

Although the aforementioned symmetry criterion results in valuable design insights, it should be kept in mind that the most accurate criterion for minimizing upconversion of low frequency noise is minimizing the dc value of the effective ISF to be defined by (4.34) in Section 4.5. In certain cases, such as the one shown in Figure 4.33, the most

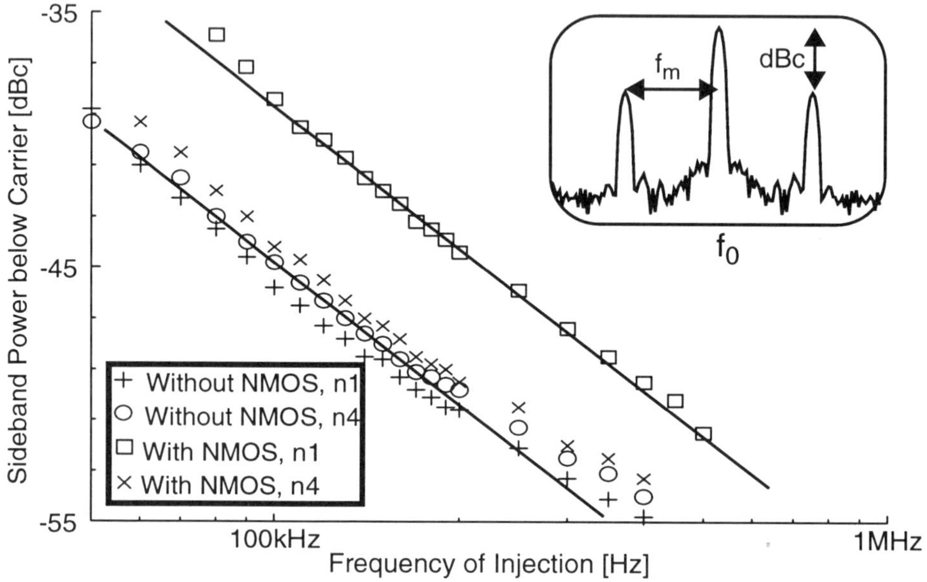

FIGURE 4.34 Power of sidebands caused by low frequency injection into symmetric and asymmetric nodes of the ring oscillator of Figure 4.35.

symmetric rise and fall time does not necessarily correspond to minimum upconversion, due to time varying capacitors and cyclostationarity of noise sources. In fact, the optimum W_p to W_n ratio in the particular example of Figure 4.33 is seen to differ considerably from that used in conventional ring oscillator designs.

The foregoing simulation results can also be verified experimentally. Figure 4.34 shows the results of an experiment performed on a 5-stage single-ended ring oscillator in which one of the stages is modified to allow the addition of an extra pulldown NMOS transistor, as shown in Figure 4.35. A sinusoidal current of 20μA (rms) is injected into node *n1* with and without the extra pulldown transistor. For comparison, this experiment is repeated for node *n4* of the oscillator before and after adding the extra pulldown transistor. Note that in Figure 4.34 the sideband power is 7dB larger when noise is injected into the node with the asymmetrical waveform, while the sidebands due to signal injection at the symmetric nodes are essentially unchanged with the modification.

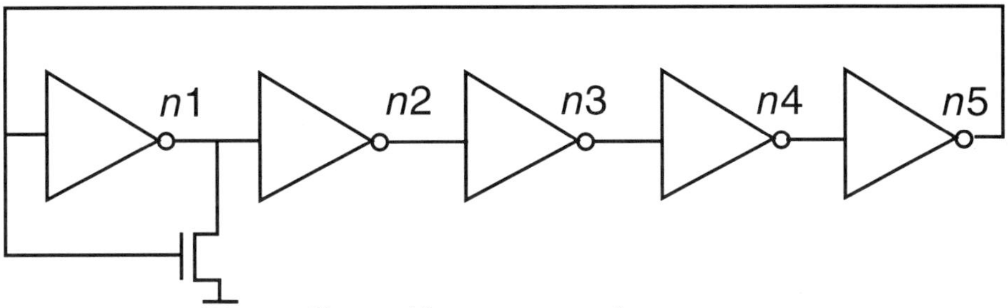

FIGURE 4.35 Ring oscillator with one asymmetric stage.

4.5 Other Time-Variant Effects

In addition to the periodically time-varying nature of the system itself, there can be other complications due to the presence of periodically time-varying noise sources, voltage-dependent capacitors or current-dependent inductors. Section 4.5.1 deals with the modeling of cyclostationary noise sources, while the effect of voltage/current-dependent elements is modeled in Section 4.5.2.

4.5.1 Cyclostationary Noise Sources

In practical oscillators, the statistical properties of some of the random noise sources may change with time in a periodic manner. These sources are referred to as cyclostationary. For instance, the channel noise of a MOS transistor in an oscillator is cyclostationary because the noise power is modulated by the gate-source overdrive which varies with time periodically. There may be other noise sources in the circuit whose statistical properties do not depend on time and the operation point of the circuit, and are therefore called stationary. Thermal noise of a resistor is an example of a stationary noise source.

These concepts can be understood best in the context of an example. Consider the Colpitts oscillator of Figure 4.9. The simulated collector voltage and current of the

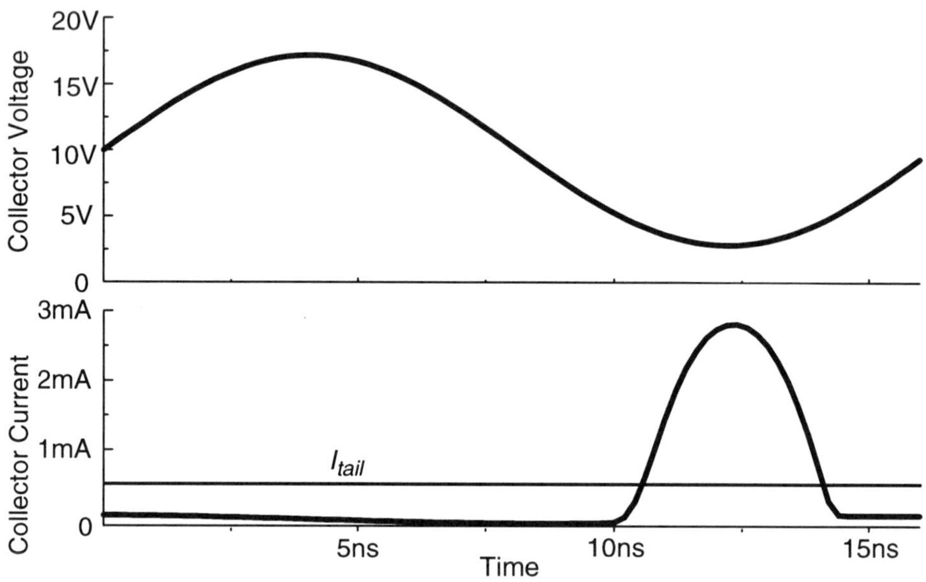

FIGURE 4.36 Collector voltage and collector current of the Colpitts oscillator of Figure 4.9.

transistor are shown in Figure 4.36. Note that the collector current consists of a short period of large current followed by a quiet interval[1]. This behavior is investigated in more details in Appendix F. The power of collector shot noise is proportional to the instantaneous collector current of the transistor[2]; therefore, it has the maximum power during the peak of collector current. Figure 4.37 shows one sample of collector shot noise of the bipolar transistor.

A white cyclostationary noise current $i_n(t)$ can always be decomposed as

$$i_n(t) = i_{n0}(t) \cdot \alpha(\omega_0 t) \qquad \textbf{(4.28)}$$

1. The reason for this behavior is investigated further in Appendix F.

2. The reason that the collector shot noise in bipolar transistors and the drain thermal noise in MOS transistors can be modeled this way has its roots in the physics of the device. In general, the expressions for the noise in the transistor are valid only after electrons and holes reach thermal equilibrium; however, in most practical applications, thermal equilibrium is reached much faster than the maximum frequency of operation of the transistor [140].

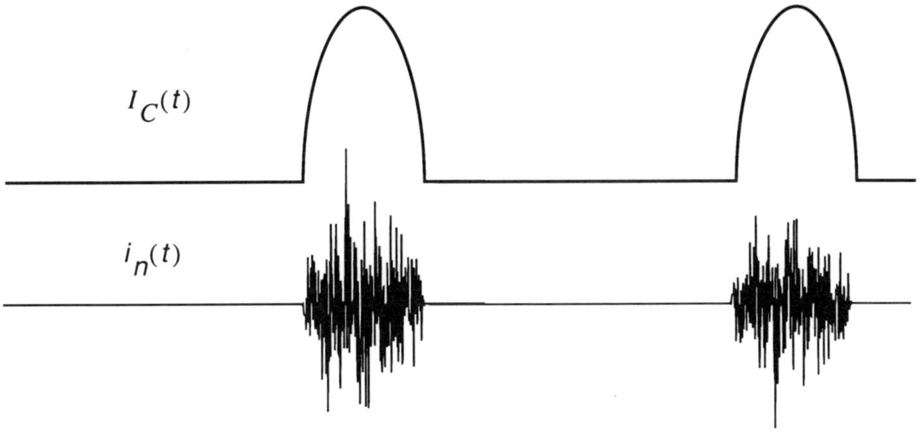

FIGURE 4.37 Collector current and a sample of collector shot noise.

where $i_{n0}(t)$ is a white *stationary* process and $\alpha(\omega 0t)$ is a deterministic periodic function describing the noise amplitude modulation and therefore is referred to as the *noise modulating function* (NMF). The NMF, $\alpha(\omega 0t)$, is normalized to a maximum value of 1. This way, $\overline{i_{n0}^2}/\Delta f$ is equal to the maximum of the periodically varying noise power density, $\overline{i_n^2(t)}/\Delta f$. Applying (4.28) to (4.4), $\phi(t)$ may be rewritten as:

$$\phi(t) = \int_{-\infty}^{t} i_n(\tau)\frac{\Gamma(\omega_0\tau)}{q_{max}}d\tau = \int_{-\infty}^{t} i_{n0}(\tau)\frac{\alpha(\omega_0\tau)\Gamma(\omega_0\tau)}{q_{max}}d\tau \tag{4.29}$$

As can be seen, cyclostationary noise can be treated as a stationary noise applied to a system with a new ISF given by[1]

$$\Gamma_{NMF}(x) = \Gamma(x)\cdot\alpha(x) \tag{4.30}$$

where $\alpha(\omega_0 t)$ can be derived easily from device noise characteristics and the noiseless steady-state waveform. Note that there is a strong correlation between the cyclostationary noise source and the waveform of the oscillator. The maximum of the noise

1. As mentioned earlier the ISF in the argument of (4.4) should appear as $\Gamma[\omega_0 t+\phi(t)]$, however, the NMF is shifted by the same amount and therefore appears in (4.29) as $\alpha[\omega_0 t+\phi(t)]$. Therefore, (4.30) holds at all times.

power always recurs at a certain point of the oscillatory waveform, thus the average of the noise may not be a good representation of the noise power.

Also note that as the waveform deviates in time from the noiseless waveform due to phase noise, $\alpha(\omega_0 t)$ shifts by exactly the same amount because the noise sources are modulated by the oscillator voltages and currents. Therefore, they will always have a constant phase relationship and (4.30) will be valid at all times. The relative timing of the cyclostationary noise sources with respect to the impulse sensitivity function can drastically change the effect of those noise sources.

Two examples are considered to provide some design insight into the effect of cyclostationary noise sources. As a first example consider the Colpitts oscillator of Figure 4.9. As can be seen in Figure 4.36, the surge of collector current occurs at the minimum of the voltage across the tank, where the ISF is small. The collector shot noise has its maximum power when the collector current is maximum, as shown in Figure 4.37. This fortunate coincidence lowers the phase noise degradation due to the collector shot noise, because the maximum noise power always coincides with the minimum phase noise sensitivity. This concept can be more accurately described using the effective ISF defined by (4.30).

Functions $\Gamma(x)$, $\alpha(x)$ and $\Gamma_{NMF}(x)$ for this oscillator are shown in Figure 4.38. Note that, in this case, $\Gamma_{NMF}(x)$ has a much smaller rms value than $\Gamma(x)$, and hence the effect of cyclostationarity is very significant for the LC oscillator and cannot be neglected.

This observation suggests that it is highly desirable for the cyclostationary noise sources to have their maximum power at the minimum sensitivity point. All other parameters being similar, the designer should seek topologies that result in a minimum value for the effective ISF.

The second example is the inverter-chain ring oscillator of Figure 4.6. The situation is quite different for the ring oscillator because the devices have an unfortunate coincidence of maximum current, maximum $\Gamma(x)$, and maximum noise power. Functions $\Gamma(x)$, $\alpha(x)$ and $\Gamma_{NMF}(x)$ for the ring oscillator of Figure 4.6 are shown in Figure 4.39.

Note that in the case of the ring oscillator $\Gamma(x)$ and $\Gamma_{NMF}(x)$ are almost identical. This unfortunate situation is one of the reasons why ring oscillators in general have inferior phase noise performance compared to a Colpitts LC oscillator[1].

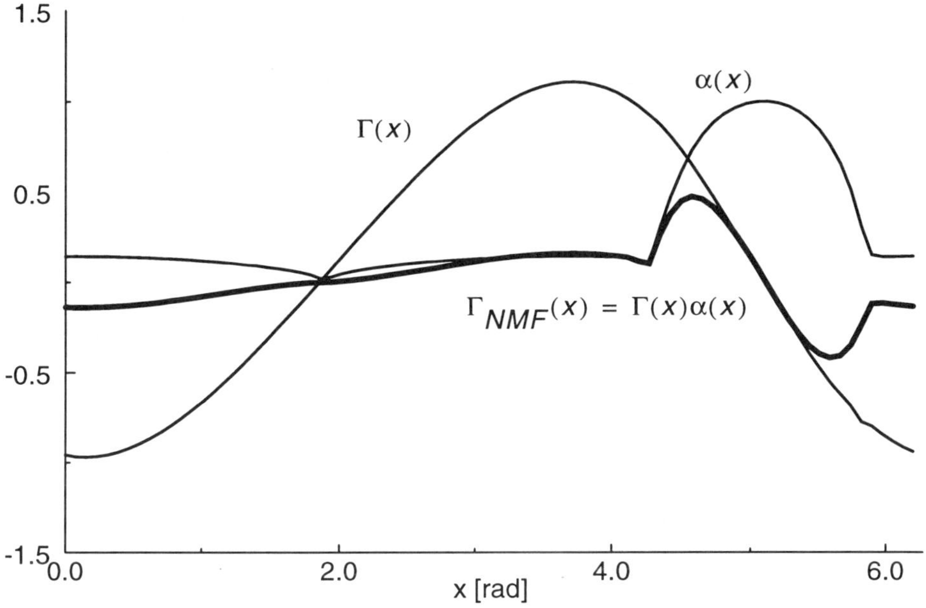

FIGURE 4.38 $\Gamma(x)$, $\alpha(x)$ and $\Gamma_{NMF}(x)$ for the Colpitts oscillator of Figure 4.9.

The foregoing observation indicates that the cyclostationary properties of noise are less important in the treatment of phase noise of ring oscillators. This lack of strong dependence was also shown through direct simulation for differential ring oscillators in [47].

4.5.2 Voltage Dependent Capacitors

Many capacitors in practical oscillators are voltage dependent, such as junction and gate capacitors. Since, in an oscillator, the node voltages vary in a periodic manner, such a capacitor will be periodically time varying and therefore

1. There is another important reason for the inferior phase noise performance of ring oscillators. A ring oscillator stores a certain amount of energy on the capacitors during every cycle and then dissipates all the stored energy during the same cycle, while an LC resonator dissipates only $2\pi/Q$ of the total stored energy during one cycle. Thus, for a given power dissipation in steady state, a ring oscillator suffers from a smaller maximum charge swing, q_{max}.

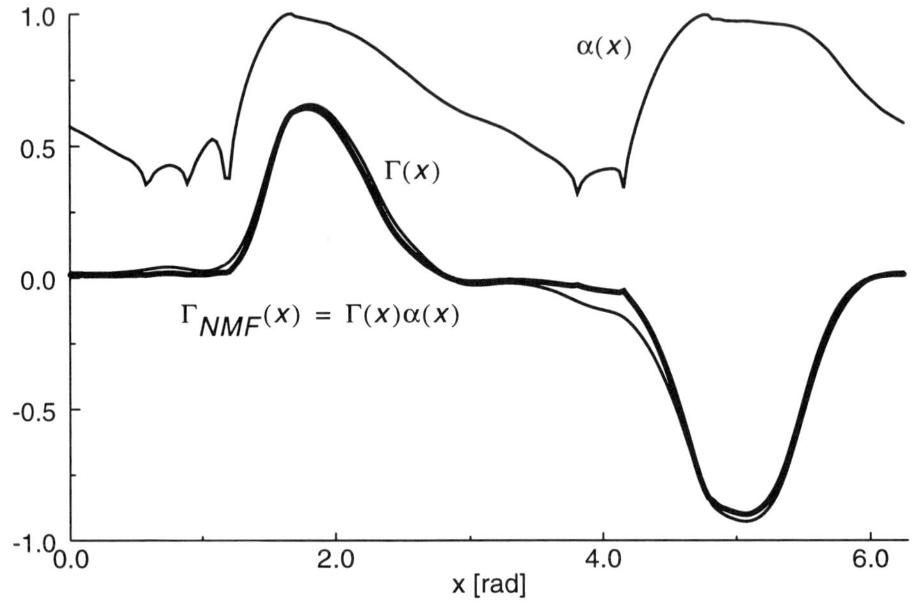

FIGURE 4.39 $\Gamma(x)$, $\alpha(x)$ and $\Gamma_{NMF}(x)$ for the ring oscillator of Figure 4.6.

$$C(\omega_0 t) = C_{max} \cdot \beta(\omega_0 t) \tag{4.31}$$

where C_{max} is the maximum value of the capacitor and $\beta(x)$ is a periodic unitless function with a maximum of 1 and a period of 2π. Function $\beta(x)$ will be referred to as the *capacitor modulating function* (CMF). Applying (4.31) to (4.4), $\phi(t)$ will be given by

$$\phi(t) = \int_{-\infty}^{t} i_n(\tau) \frac{\Gamma(\omega_0 \tau)}{V_{max} C(\omega_0 \tau)} d\tau = \int_{-\infty}^{t} i_{n0}(\tau) \frac{\Gamma(\omega_0 \tau)}{q_{max} \cdot \beta(\omega_0 \tau)} d\tau \tag{4.32}$$

Therefore, such a capacitor can be modeled as a time-invariant capacitor with a value C_{max}, along with a new ISF given by

$$\Gamma_{CMF}(x) = \frac{\Gamma(x)}{\beta(x)} \tag{4.33}$$

Note that $\beta(x)$ can be obtained easily from steady-state simulation of the oscillator. A similar approach can be taken to model current dependent inductors[1].

The effect of cyclostationary noise sources and periodically time-varying capacitors can be lumped into an effective ISF defined as

$$\Gamma_{eff}(x) = \Gamma(x) \cdot \frac{\alpha(x)}{\beta(x)} \tag{4.34}$$

Hence, in order to take these time-variant effects into account, the effective ISF defined by (4.34) should be used in all subsequent calculations. These concepts will be generalized further in CHAPTER 7 to accommodate multiple noise sources.

4.6 Amplitude Response

Up to this point, the effect of amplitude fluctuations has been ignored. As will be seen shortly, while the close-in sidebands are almost always dominated by phase noise, the far-out sidebands are significantly affected by amplitude noise. In this section, a simple model for analyzing the consequences of amplitude fluctuations is presented.

As mentioned in Section 4.1, unlike the induced excess phase that persists indefinitely, the excess amplitude, $A(t)$, due to a current impulse decays with time. This decay is the direct result of the amplitude restoring mechanisms always present in practical oscillators. Depending on the oscillator and the nature of the amplitude limiting mechanism, the excess amplitude may decay very slowly (*e.g.*, in a harmonic oscillator with a high quality resonant circuit) or very quickly (*e.g.*, the ring oscillator of Figure 4.6). Also in some oscillators, there may be ringing in the output amplitude, as shown in Figure 4.40.

A current impulse with an area, Δq, will induce an instantaneous change in the capacitor voltage given by (4.1). This, in turn, will result in a change in the oscillator amplitude that depends on the instant of injection, as shown in Figure 4.2. The amplitude change is proportional to the instantaneous normalized voltage change[2], $\Delta V / V_{max}$, for small injected charge, *i.e.*,

1. An inductor with a ferromagnetic core whose relative magnetic permeability, μ_{rel}, is a function of magnetic intensity, H, is an example of a current-dependent inductor.

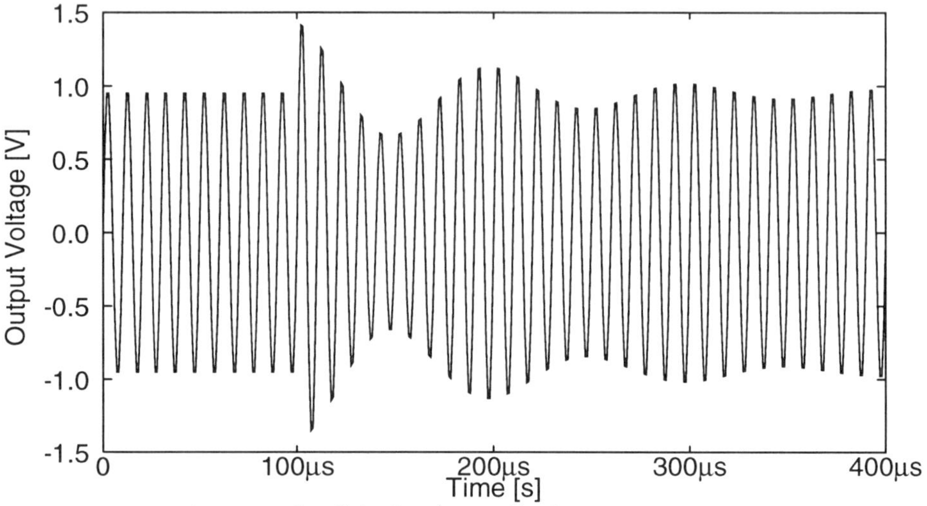

FIGURE 4.40 An example of ringing in amplitude response.

$$\Delta A = \Lambda(\omega_0 t)\frac{\Delta V}{V_{max}} = \Lambda(\omega_0 t)\frac{\Delta q}{q_{max}} \qquad \Delta q \ll q_{swing} \tag{4.35}$$

where $\Lambda(\omega_0 t)$ is a periodic function that determines the sensitivity of each point on the waveform to an impulse and is called the *amplitude impulse sensitivity function*. It is the amplitude counterpart of the phase impulse sensitivity function, $\Gamma(\omega_0 t)$. From a development similar to that of Section 4.1, the amplitude impulse response can be written as

$$h_A(t, \tau) = \frac{\Lambda(\omega_0 t)}{q_{max}} d(t-\tau) \tag{4.36}$$

where $d(t\text{-}\tau)$ is a function that defines how the excess amplitude decays. Figure 4.41 shows two hypothetical examples of $d(t)$ for a low Q oscillator with overdamped response and a high Q oscillator with underdamped amplitude response.

In general, the exact amplitude response depends on the details of the particular oscillator of interest. For most oscillators, the amplitude limiting system can be approxi-

2. Note that the amplitude change is only equal to the normalized voltage change $\Delta V/V$ if the impulse is injected at the peak of the sinusoidal tank voltage.

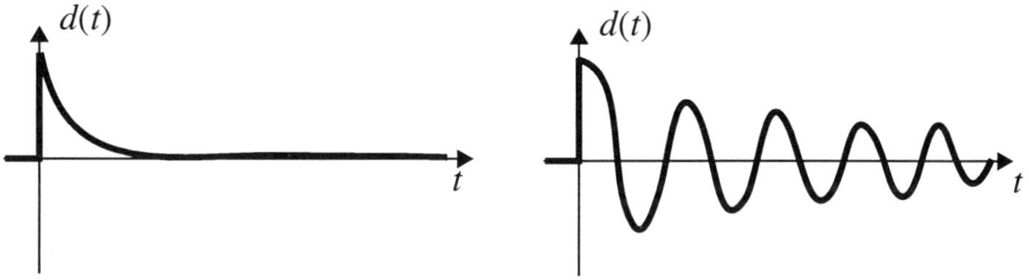

FIGURE 4.41 Overdamped and underdamped amplitude responses.

mated as first or second order [10]. The function $d(t-\tau)$ typically will thus be either a dying exponential or a damped sinusoidal as shown in Figure 4.41.

A first order analysis to determine the amplitude response of a harmonic oscillator can be insightful. In a resonator, Q is defined as

$$Q = 2\pi \frac{E_{stored}}{E_{diss}} \tag{4.37}$$

where E_{stored} is the total energy stored in the resonator and E_{diss} is the energy dissipated every cycle due to losses in the tank. During each cycle, the active device injects E_{active} energy into the tank. Therefore, the rate of change in the total energy is

$$\frac{dE}{dt} = \frac{E_{active} - E_{diss}}{T} = \frac{E_{active}}{T} - \frac{2\pi}{T}\frac{E}{Q} \tag{4.38}$$

where $E(t)$ is the total tank energy at time t. In steady state, E stays constant and, therefore, $E_{active} = E_{diss}$.

The total energy can be expressed as $E = E_{steady} + \Delta E$, where E_{steady} is the steady-state tank energy and ΔE is the change in the energy. By rewriting E this way and assuming that E_{active} does not change, (4.38) will result in

$$\frac{d(\Delta E)}{dt} = -\frac{2\pi}{T}\frac{\Delta E}{Q} = -\frac{\omega_0}{Q}\Delta E \tag{4.39}$$

whose solution is a decaying exponential:

$$\Delta E(t) = \Delta E_0 \cdot e^{-\omega_0 \tau / Q} \cdot u(t-\tau) \tag{4.40}$$

where ΔE_0 is the initial excess energy of the tank. In the case of an LC tank, the total energy is

$$E = \frac{C}{2}(V_{max} + \Delta V)^2 = \frac{C}{2}V_{max}^2 + CV_{max}\Delta V + \frac{C}{2}\Delta V^2 \tag{4.41}$$

where C is the tank capacitor and ΔV is the induced voltage change. If the induced voltage change is small compared to the steady-state voltage swing, the last term in (4.41) can be neglected. Therefore,

$$\Delta E \approx CV_{max}\Delta V \tag{4.42}$$

The excess amplitude will be given by

$$\Delta A(t) = \Delta A_0 \cdot e^{-\omega_0 \tau / Q} \cdot u(t - \tau) \tag{4.43}$$

and accordingly

$$d(t - \tau) = e^{-\omega_0 (t - \tau)/Q} \cdot u(t - \tau) \tag{4.44}$$

Using (4.36) and (4.44), the excess amplitude response to an arbitrary input current, $i(t)$, can be calculated using the superposition integral,

$$A(t) = \int_{-\infty}^{t} \frac{i(\tau)}{q_{max}} \Lambda(\omega_0 \tau) e^{-\omega_0 (t - \tau)/Q} d\tau \tag{4.45}$$

If $i(t)$ is a white noise source with power spectral density $\overline{i_n^2}/\Delta f$, the output power spectrum of the amplitude noise, $A(t)$, can be shown to be

$$\mathcal{L}_{amplitude}\{\Delta\omega\} = \frac{\Lambda_{rms}^2}{q_{max}^2} \cdot \frac{\overline{i_n^2}/\Delta f}{2 \cdot \left(\frac{\omega_0^2}{Q^2} + \Delta\omega^2\right)} \tag{4.46}$$

where Λ_{rms} is the rms value of $\Lambda(\omega_0 t)$. As can be seen, the output amplitude noise spectrum is a Lorentzian, as depicted in Figure 4.42. If $\mathcal{L}_{total}$ is measured, the sum of both $\mathcal{L}_{amplitude}$ and $\mathcal{L}_{phase}$ will be observed and hence there will be a pedestal in the phase noise spectrum at ω_0/Q as shown in Figure 4.42. Also note that the significance

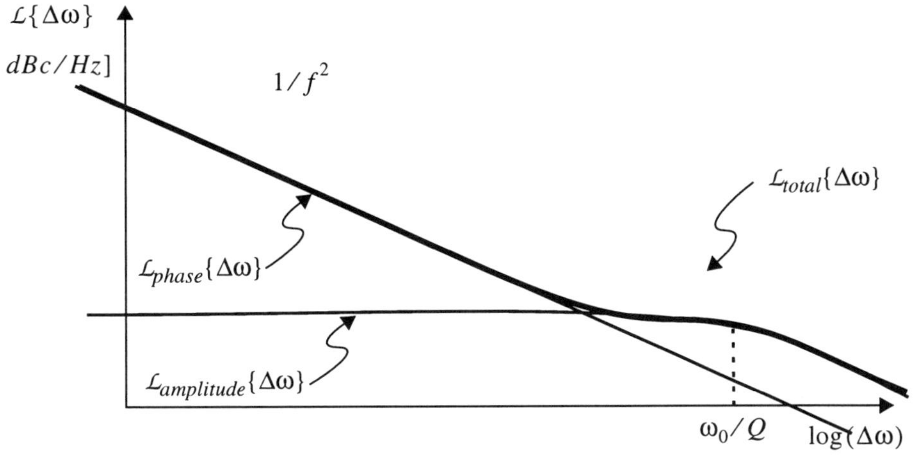

FIGURE 4.42 Phase, amplitude and total output sideband powers for the overdamped exponentially decaying amplitude response.

of the amplitude response depends greatly on Λ_{rms} which, in turn, depends on the topology.

In the case of an underdamped response, $d(t\text{-}\tau)$ will have a sinusoidal term, *i.e.*,

$$d(t-\tau) = e^{-\omega_0(t-\tau)/Q} \cdot \cos[\omega_1(t-\tau)] \cdot u(t-\tau) \tag{4.47}$$

where ω_1 is the natural frequency of the amplitude response. In this case, there will be peaking in the amplitude frequency response as shown in Figure 4.43.

4.7 Relationship to Previous Models

The time-varying model proposed here reduces to earlier models if approximate simplifying assumptions are made. In particular, consider models which assume that linear time-invariance holds, that all noise sources are stationary, that only the noise in the vicinity of ω_0 is important, and that the noise-free waveform is a perfect sinusoid [22][25][37]. These assumptions are equivalent to discarding all but the c_1 term in the ISF and setting c_1=1. As a specific example, consider the oscillator of Figure 3.2. The

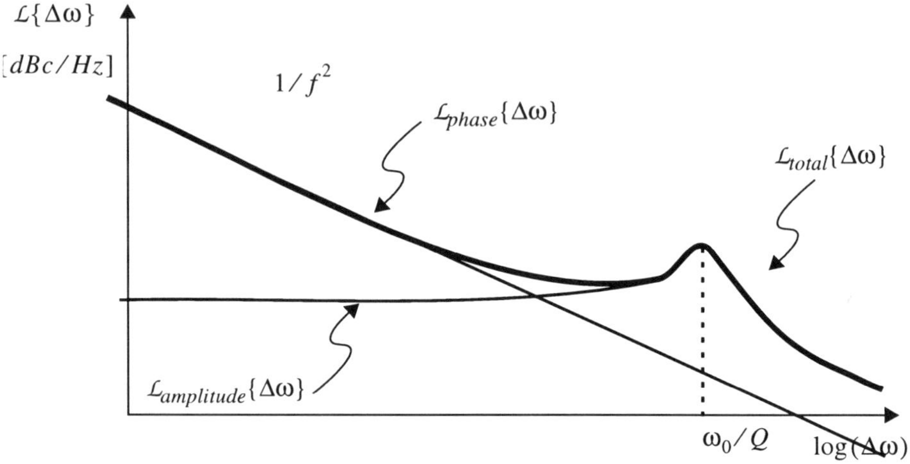

FIGURE 4.43 Phase, amplitude and total output sideband powers for the underdamped exponentially decaying amplitude response

phase noise due solely to the tank equivalent parallel conductance, G_L, can be found by applying the following to (4.14):

$$\overline{i_n^2}/\Delta f = 4kTG_L$$
$$q_{max} = C \cdot V_{max} \tag{4.48}$$

where C is the tank capacitor, and V_{max} is the maximum voltage swing across the tank. Equation (4.14) thus reduces to

$$\mathcal{L}\{\Delta\omega\} = \frac{kT}{V_{max}^2} \cdot \frac{G_L}{(C\omega_0)^2} \cdot \left(\frac{\omega_0}{\Delta\omega}\right)^2 \tag{4.49}$$

Since [37] assumes equal contributions from amplitude and phase portions to $\mathcal{L}_{total}\{\Delta\omega\}$, the result obtained in [37] is two times larger than the result of (4.49).

Equations (4.49) and (4.26) result in

$$\mathcal{L}\{\Delta\omega\} = \frac{4\Gamma_{rms}kT}{P_s} \cdot \left[1 + \left(\frac{\omega_0}{2Q\Delta\omega}\right)^2\right] \cdot \left(1 + \frac{c_0^2}{\Gamma_{rms}^2} \cdot \frac{\omega_{1/f}}{|\Delta\omega|}\right) \tag{4.50}$$

Note that (4.50) can be used to calculate the fitting parameters used in (3.1), (F, and $\Delta\omega_{1/f^3}$) in terms of the dc and rms values of the ISF and device $1/f$ noise corner, $\omega_{1/f}$. Assuming that the total noise contribution in an oscillator with a parallel tank can be modeled using an excess noise number [22], F, the fitting parameters in (3.1) are

$$F = 2\Gamma_{rms}$$

$$\omega_{1/f^3} = \omega_{1/f} \cdot \frac{c_0^2}{\Gamma_{rms}^2} \tag{4.51}$$

It is important to note that the $1/f^2$-to-noise floor corner predicted by (4.50) does not necessarily coincide with the actual corner, as can be seen in Figure 4.42. More general expressions for the case of multiple noise sources will be developed in CHAPTER 7.

4.8 Summary

The phase impulse response of an oscillator was defined using the impulse sensitivity function (ISF). The response of an oscillator to deterministic and random noise sources was calculated in terms of Fourier coefficients of the ISF. Expressions for the phase noise in $1/f^2$ and $1/f^3$ regions, as well as for timing jitter, were obtained. The effect of symmetry on the upconversion of low frequency noise was shown, and time-variant effects and their modeling were investigated. The output spectrum due to amplitude noise was calculated.

CHAPTER 5

Jitter and Phase Noise in Ring Oscillators

Due to their amenability to integration, ring oscillators have become an essential building block in many digital and communication systems. They are used as voltage controlled oscillators (VCOs) in applications such as clock recovery circuits for serial data communications [110]-[113], disk drive read channels [114][115], on-chip clock distribution [116]-[119], and integrated frequency synthesizers [119][120]. In Section 5.1, a closed form expression for the rms and dc values of the ISF for ring oscillators is derived. These approximate rms and dc values are used to obtain closed-form expressions for phase noise and jitter in ring oscillators in Section 5.2. The effect of correlated noise sources is investigated in Section 5.3. Design implications such as the question of single-ended vs. differential implementation of ring oscillators and the optimum number of stages are addressed in Section 5.4. Finally, these expressions are verified experimentally for various single-ended and differential ring oscillators in Section 5.5.

5.1 The Impulse Sensitivity Function for Ring Oscillators

To calculate phase noise and jitter using (4.16) and (4.23), one needs to know the dc and rms values of the ISF. In this section, approximate closed-form equations for the dc and rms values of the ISF of ring oscillators are obtained. The special case of equal

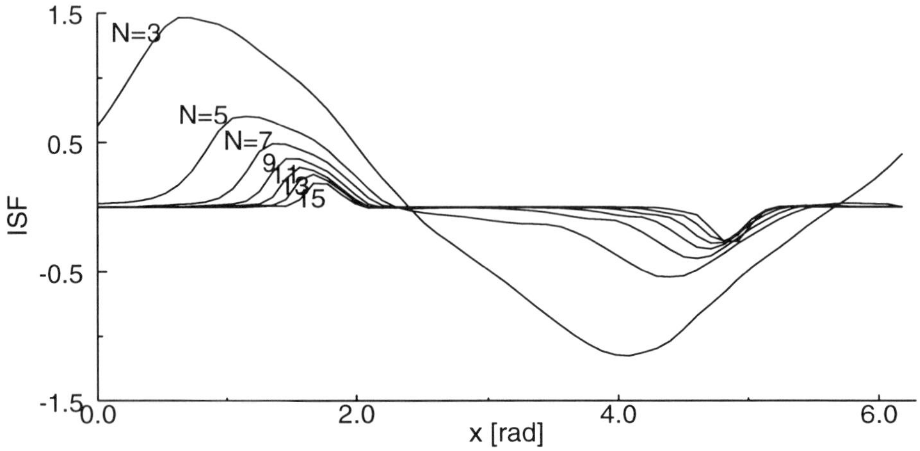

FIGURE 5.1 The ISF for ring oscillators of the same frequency with different number of stages.

rise and fall time is analyzed in Section 5.1.1, while Section 5.1.2 investigates the more general case of unequal rise and fall times.

5.1.1 Equal Rise and Fall Times

It is instructive to look at the actual ISF of ring oscillators to gain insight into what constitutes a good approximation. Figure 5.1 shows the shape of the ISF for a group of single-ended CMOS ring oscillators. The frequency of oscillation is kept constant (through adjustment of channel length), while the number of stages is varied from 3 to 15 (in odd numbers). The ISF is calculated using the first method presented in Appendix D.

As can be seen, increasing the number of stages reduces the peak value of the ISF. The reason is that the transitions of the normalized waveform become faster for larger N in this constant-frequency scenario. Since the sensitivity is inversely proportional to the slope, the peak of the ISF drops. Also the widths of the lobes of the ISF decrease as N becomes larger since each transition occupies a smaller fraction of the period. Based on these observations, the ISF of ring oscillators with equal rise and fall times can be approximated as two identical triangles, as shown in Figure 5.2.

The ISF has a maximum of $1/f'_{max}$, where f'_{max} is the maximum slope of the normalized waveform f in (2.1). Also the width of the triangles is approximately $2/f'_{max}$

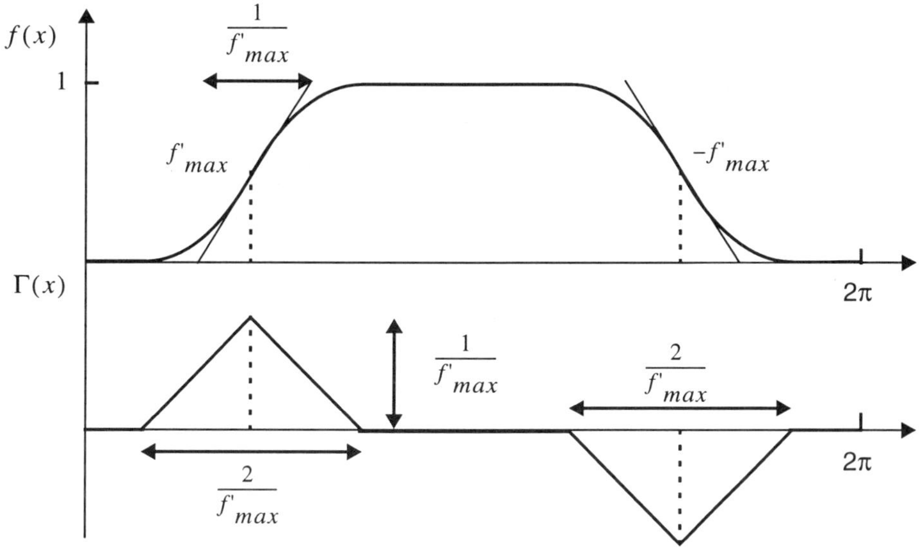

FIGURE 5.2 Approximate waveform and ISF for a ring oscillator.

and hence the slopes of the sides of the triangles are ± 1. Therefore, assuming equality of the rise- and fall-times, Γ_{rms} can be estimated as

$$\Gamma_{rms}^2 = \frac{1}{2\pi}\int_0^{2\pi}\Gamma^2(x)dx = \frac{4}{2\pi}\int_0^{1/f'_{max}}x^2dx = \frac{2}{3\pi}\left(\frac{1}{f'_{max}}\right)^3 \quad \textbf{(5.1)}$$

On the other hand, stage delay is proportional to the rise time:

$$\hat{t}_D = \frac{\eta}{f'_{max}} \quad \textbf{(5.2)}$$

where $\hat{t}_D$ is the stage delay normalized to the period and η is a proportionality constant, which is typically close to 1, as can be seen in Figure 5.3.

The period is $2N$ times longer than a single stage delay, *i.e.,*

$$2\pi = 2N\hat{t}_D = \frac{2N\eta}{f'_{max}} \quad \textbf{(5.3)}$$

Using (5.1) and (5.3), the following approximate expression for Γ_{rms} is obtained:

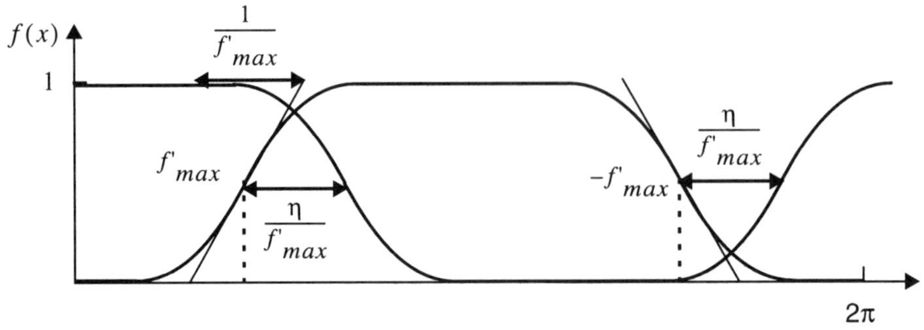

FIGURE 5.3 The relationship between risetime and delay.

$$\Gamma_{rms} \approx \sqrt{\frac{2\pi^2}{3\eta^3}} \cdot \frac{1}{N^{1.5}} \quad \textbf{(5.4)}$$

Note that the $1/N^{1.5}$ dependence of Γ_{rms} is independent of the value of η. Figure 5.4 illustrates Γ_{rms} vs. the number of stages for the ISFs shown in (5.1)Figure 5.1 with plus signs on *log-log* axes. The solid line shows the line of $\Gamma_{rms} \approx 4/N^{1.5}$, which is obtained from (5.4) for $\eta = 0.75$. To verify the generality of (5.4), a second set of simulations was performed in which a fixed channel length is maintained for all the devices in the inverters while varying the number of stages to allow different frequencies of oscillation. Again, $\Gamma(x)$ is directly simulated and its rms value is plotted in Figure 5.4 with circles. This simulation is repeated with a different supply voltage (3V as opposed to 5V) and the result is shown with crosses. As can be seen, the values of Γ_{rms} are almost identical for these three cases.

It should not be surprising that Γ_{rms} is primarily a function of N because the effect of variations in other parameters, such as q_{max} and device noise, have already been decoupled from $\Gamma(x)$; the ISF is a unitless, frequency- and amplitude-independent function.

Equation (5.4) is valid for differential ring oscillators as well, since in its derivation no assumption specific to single-ended oscillators was made. Figure 5.5 shows Γ_{rms} for three sets of differential ring oscillators, with a varying number of stages (4 to 16). The data shown with plus signs correspond to oscillators in which the total power dissipation and drain voltage swing are kept constant by scaling the tail current sources and load resistors as N changes. Members of the second set of oscillators have a fixed total power dissipation and fixed load resistors, which result in variable swings, and for whom data are shown with circles. The third case is that of a fixed tail current for

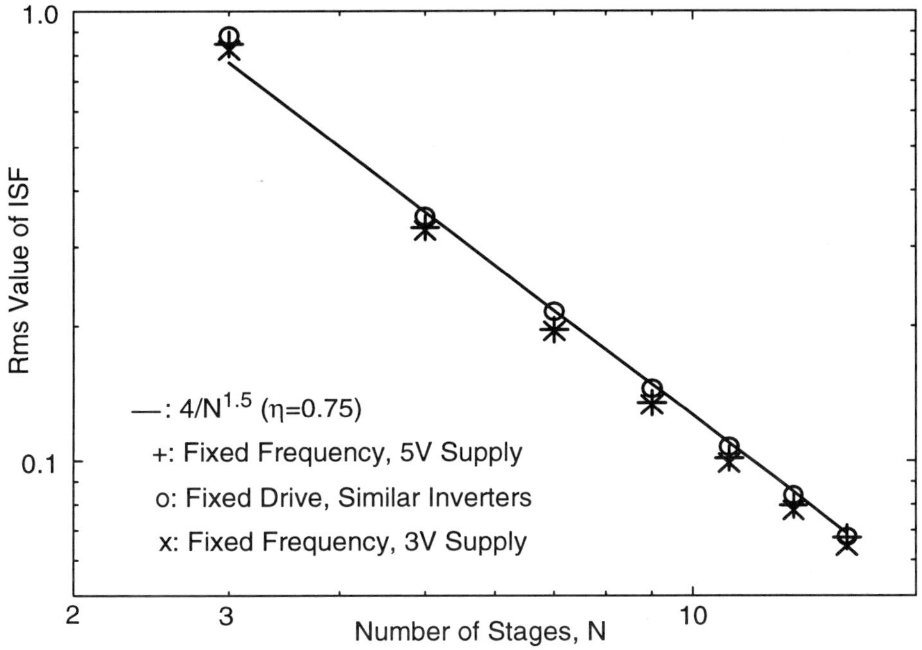

FIGURE 5.4 Rms values of the ISFs for various single-ended ring oscillators vs. number of stages.

each stage and constant load resistors, whose data are illustrated using crosses. Again, in spite of the diverse variations of frequency and other circuit parameters, the $1/N^{1.5}$ dependency of Γ_{rms} and its independence from other circuit parameters still holds. In the case of a differential ring oscillator, $\Gamma_{rms} \approx 3/N^{1.5}$, which corresponds to $\eta = 0.9$, is the best fit approximation for Γ_{rms}. This is shown with the solid line in Figure 5.5. ? Here η could easily be measured rather than best fit!

Although Γ_{rms} decreases as the number of stages increases, one should not prematurely conclude that the phase noise can be reduced using a larger number of stages, because the number of noise sources, as well as their magnitudes, also increases, for a given total power dissipation and frequency of oscillation. The question of optimal number of stages is therefore a bit involved, and will be addressed in Section 5.2.

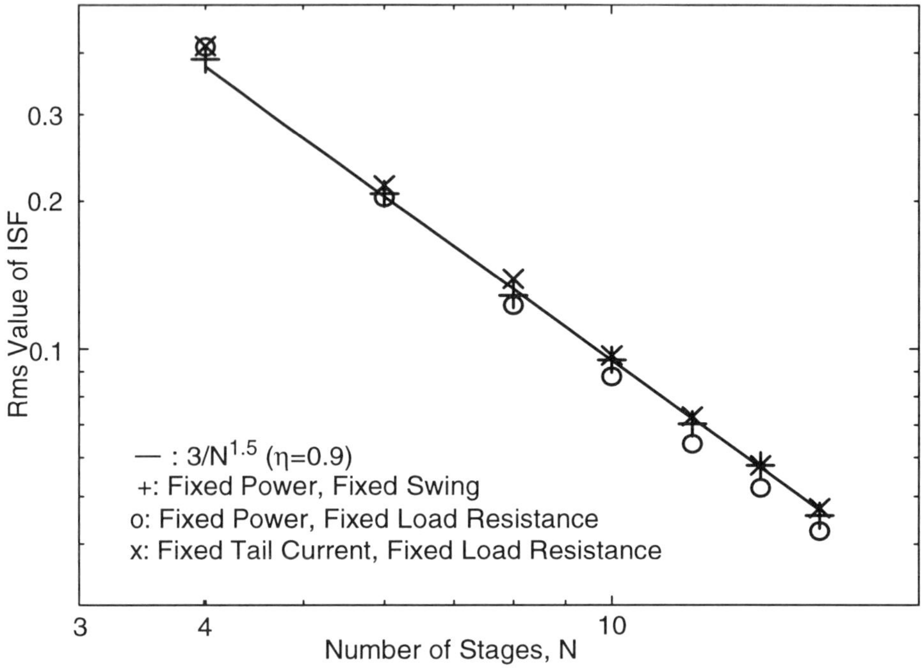

FIGURE 5.5 Rms values of the ISFs for various differential ring oscillators vs. number of stages.

5.1.2 Unequal Rising and Falling Times

For a ring oscillator with asymmetric rising and falling edges, the ISF is approximated by the function depicted in Figure 5.6.

The rms value of the approximate ISF in Figure 5.6 can be calculated as follows:

$$\Gamma_{rms}^2 = \frac{1}{\pi}\left[\int_0^{1/f'_{rise}} x^2 dx + \int_0^{1/f'_{fall}} x^2 dx\right] = \frac{1}{3\pi}\left(\frac{1}{f'_{rise}}\right)^3 (1 + A^3) \tag{5.5}$$

where f'_{rise} and f'_{fall} are the maximum slope during the rising and falling edge, respectively, and A represents the asymmetry of the waveform and is defined as

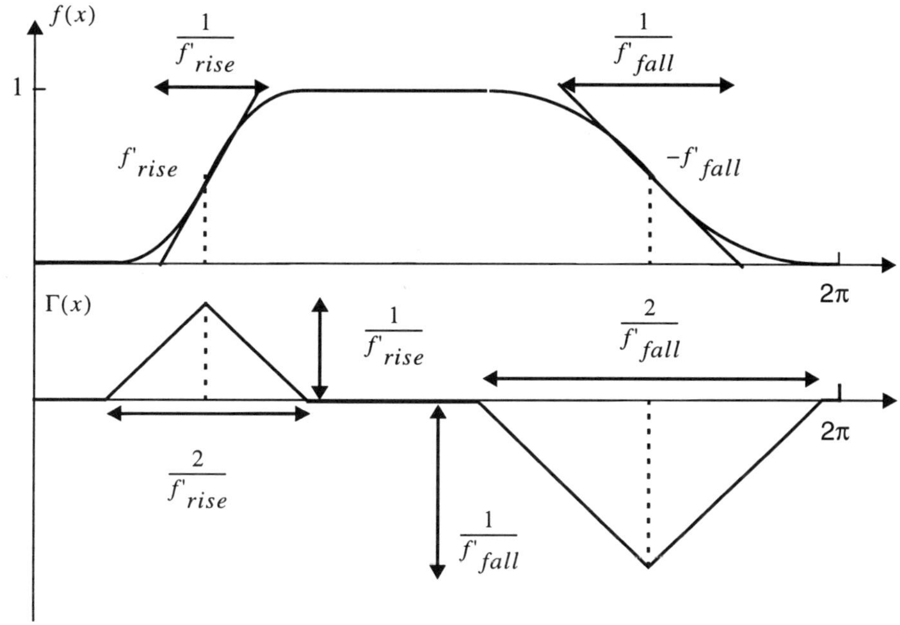

FIGURE 5.6 Approximate waveform and the ISF for asymmetric rising and falling edge. edges

$$A \equiv \frac{f'_{rise}}{f'_{fall}} \tag{5.6}$$

The period consists of N rising-edge and N falling-edge delays, *i.e.*,

$$2\pi = \eta N\left(\frac{1}{f'_{rise}} + \frac{1}{f'_{fall}}\right) = \frac{\eta N}{f'_{rise}}(1 + A) \tag{5.7}$$

Combing (5.5) and (5.7) results in the following:

$$\Gamma^2_{rms} = \frac{8\pi^2}{3\eta^3}\frac{1}{N^3}\left[\frac{1 + A^3}{(1 + A)^3}\right] \tag{5.8}$$

which reduces to (5.4) in the special case of $A = 1$, *i.e.,* symmetric rising and falling edges. The dc value of the ISF, Γ_{dc}, can be calculated from Figure 5.6, in a similar manner and is given by

$$\Gamma_{dc} = \frac{2\pi}{\eta^2}\frac{1}{N^2}\left(\frac{1-A}{1+A}\right) \tag{5.9}$$

Using (4.24), the $1/f^3$ corner is given by

$$f_{1/f^3} = f_{1/f} \cdot \frac{3}{2\eta N} \cdot \frac{(1-A)^2}{(1-A+A^2)} \tag{5.10}$$

As a special case, if the rise and fall time are symmetric, $A = 1$, and the $1/f^3$ corner approaches zero.

In the case of asymmetric rising and falling edges, both Γ_{rms} and Γ_{dc} will change. The $1/f^3$ corner of the phase noise spectrum is inversely proportional to the number of stages. Therefore, the $1/f^3$ corner can be reduced either by making the transitions more symmetric in terms of rise and fall times or by increasing the number of stages. Although the former always helps, the latter has important implications on the phase noise in the $1/f^2$ region, as will be shown in the next section.

5.2 Expressions for Jitter and Phase Noise in Ring Oscillators

Although the expressions obtained in the last section for the rms and dc values of the ISF can be used to calculate the phase noise, it is desirable to express phase noise and jitter in terms of design parameters such as power dissipation and frequency. In this section, several expressions for the phase noise and jitter of different types of ring oscillators are derived in terms of such parameters.

Throughout this section, it is assumed that the symmetry criteria for minimizing Γ_{dc} (and hence the upconversion of $1/f$ noise) are already met and that the jitter and phase noise of the oscillator are dominated by white noise. For CMOS transistors, the drain current noise spectral density is given by [136]

$$\frac{\overline{i_n^2}}{\Delta f} = 4kT\gamma g_{d0} = 4kT\gamma\mu C_{ox}\frac{W}{L}\Delta V \tag{5.11}$$

where g_{d0} is the zero bias drain-source conductance, μ is the mobility, C_{ox} is the gate oxide capacitance per unit area, W and L are the channel width and length, and ΔV is the gate voltage overdrive. The coefficient γ is 2/3 for long channel devices in the saturation region and typically 2-3 times greater for short-channel devices [134][136]. Equation (5.11) is valid in both short and long channel regimes, as long as an appropriate value for γ is used.

5.2.1 Single-Ended CMOS Ring Oscillators

The first case considered is a single-ended CMOS ring oscillator with equal-length NMOS and PMOS transistors. Assuming $V_{TN} = V_{TP}$, the maximum total channel noise from the NMOS and PMOS devices, when both the input and output are at $V_{DD}/2$, is given by

$$\frac{\overline{i_n^2}}{\Delta f} = \left(\frac{\overline{i_n^2}}{\Delta f}\right)_N + \left(\frac{\overline{i_n^2}}{\Delta f}\right)_P = 4kT\gamma\mu_{eff}C_{ox}\frac{W_{eff}}{L}\Delta V \tag{5.12}$$

where

$$W_{eff} = W_n + W_p \tag{5.13}$$

and

$$\mu_{eff} = \frac{\mu_n W_n + \mu_p W_p}{W_n + W_p} \tag{5.14}$$

and ΔV is the gate overdrive in the middle of the transition, *i.e.*, $\Delta V = (V_{DD}/2) - V_T$.

During one period, each node is charged to q_{max} and then discharged to zero. In an N-stage single-ended ring oscillator, the power dissipation associated with this process is $Nq_{max}V_{DD}f_0$. However, during the transitions, some extra current, known as crowbar current, is drawn from the supply. This current does not contribute to charging and discharging the capacitors since it goes directly from supply to ground through both transistors. These two components of the total current drawn from sup-

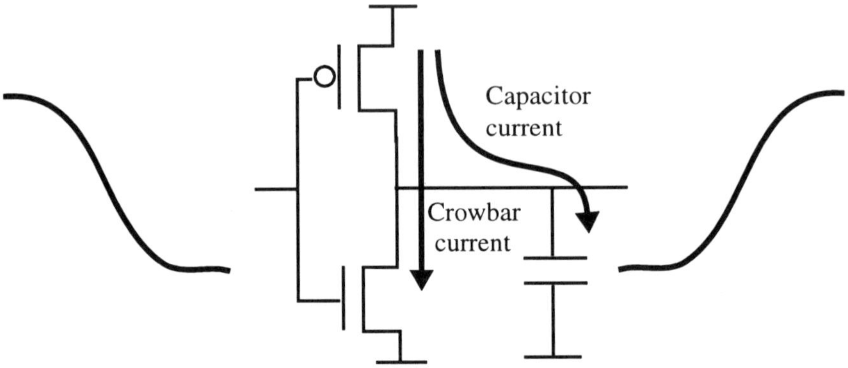

FIGURE 5.7 Capacitor and crowbar current in an inverter.

ply are shown in Figure 5.7. In a symmetric ring oscillator, these two components are comparable[1] and their difference will depend on the ratio of the rise time to stage delay; therefore, the total power dissipation is approximately given by

$$P = 2\eta N V_{DD} q_{max} f_0 \tag{5.15}$$

Assuming $\mu_n W_n = \mu_p W_p$ to make the waveforms symmetric to first order, the frequency of oscillation for long channel devices can be approximated by

$$f_0 = \frac{1}{2Nt_D} = \frac{1}{\eta N(t_r + t_f)} \approx \frac{\mu_{eff} W_{eff} C_{ox} \Delta V^2}{8\eta N L q_{max}} \tag{5.16}$$

where t_D is the delay of each stage and t_r and t_f are the rise and fall time associated with the maximum slope during a transition.

Assuming that the thermal noise sources of the different devices are uncorrelated, and assuming that the waveform (and hence the ISF) of all the nodes are the same except for a phase shift, the total phase noise due to all N noise sources is N times the value given by (4.16). Taking only these inevitable noise sources into account, (4.16), (5.4), (5.12), (5.15) and (5.16) result in the following expressions for phase noise and jitter:

1. The ratio of these two current components can change significantly with the threshold voltages, various capacitors and supply voltages. Nevertheless, the general behavior of the final result does not depend on this particular ratio.

$$\mathcal{L}\{\Delta\omega\} \approx \frac{8}{3\eta} \cdot \frac{kT}{P} \cdot \frac{V_{DD}}{V_{char}} \cdot \frac{\omega_0^2}{\Delta\omega^2} \tag{5.17}$$

$$\kappa \approx \sqrt{\frac{8}{3\eta}} \cdot \sqrt{\frac{kT}{P} \cdot \frac{V_{DD}}{V_{char}}} \tag{5.18}$$

where κ is the proportionality constant defined by (2.6) and V_{char} is the *characteristic voltage* of the device. For long channel devices $V_{char} = \Delta V / \gamma$. Any extra disturbance, such as substrate and supply noise, or noise contributed by extra circuitry or asymmetry in the waveform, will result in a larger number than (5.17) and (5.18). Note that lowering threshold voltages reduces the phase noise and could be exploited to improve the phase noise. Therefore, the minimum achievable phase noise and jitter for a single-ended CMOS ring oscillator, assuming all symmetry criteria are met, occurs for zero threshold voltage:

$$\mathcal{L}\{\Delta\omega\} > \frac{16\gamma}{3\eta} \cdot \frac{kT}{P} \cdot \frac{\omega_0^2}{\Delta\omega^2} \tag{5.19}$$

$$\kappa > \sqrt{\frac{16\gamma}{3\eta}} \cdot \sqrt{\frac{kT}{P}} \tag{5.20}$$

As can be seen, the minimum phase noise is inversely proportional to the power dissipation and grows quadratically with the oscillation frequency. Further, note the lack of dependence on the number of stages (for a given power dissipation and oscillation frequency). Evidently, the increase in the number of noise sources (and in the maximum power due to the higher transition currents required to run at the same frequency) essentially cancels the effect of decreasing Γ_{rms} as N increases, leading to no net dependence of phase noise on N. This somewhat surprising result may explain the confusion that exists regarding the optimum N since there is not a strong dependence on the number of stages for single-ended CMOS ring oscillators. Note that (5.19) and (5.20) establish lower bounds and therefore should not be used to calculate the phase noise and jitter of an arbitrary oscillator. Also in using (5.17) and (5.18) one should verify the validity of the assumptions leading to these expressions. To calculate the phase noise and jitter of an arbitrary oscillator, (4.16) and (4.23) should be used, respectively.

A similar calculation for the short channel case can be carried out. For such devices, the drain current may be expressed as

$$I_D = \frac{\mu C_{ox}}{2} W E_c \Delta V \tag{5.21}$$

where E_c is the critical electrical field[1]. Combining (5.11) with (5.21), the following expression for the drain current noise of a MOS device in short channel is obtained:

$$\frac{\overline{i_n^2}}{\Delta f} = 8kT\frac{\gamma I_D}{E_C L} \tag{5.22}$$

The frequency of oscillation can be approximated by

$$f_0 = \frac{1}{2Nt_D} = \frac{1}{\eta N(t_r + t_f)} \approx \frac{\mu_{eff} W_{eff} C_{ox} \Delta V^2}{8\eta N L q_{max}} \tag{5.23}$$

Using (5.22), the same expressions for phase noise and jitter as given by (5.17) and (5.18) are obtained, except for a new V_{char}:

$$V_{char} = \frac{E_c L}{\gamma} \tag{5.24}$$

which results in a larger phase noise and jitter than the long channel case by a factor of $\gamma \Delta V / E_c L$. As before, note the lack of dependence on the number of stages in the case of short-channel devices.

5.2.2 Differential CMOS Ring Oscillators

Now consider a differential MOS ring oscillator with resistive load. The total power dissipation is

$$P = N I_{tail} V_{DD} \tag{5.25}$$

where N is the number of stages, I_{tail} is the tail bias current of the differential pair, and V_{DD} is the supply voltage. The frequency of oscillation can be approximated by

1. Critical field is defined as the value of electric field resulting in half the carrier velocity expected from low field mobility.

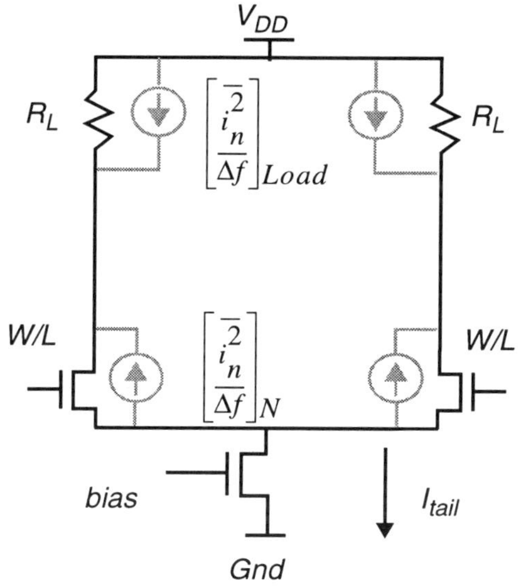

FIGURE 5.8 The noise sources in the differential buffer stage.

$$f_0 = \frac{1}{2Nt_D} \approx \frac{1}{2\eta N t_r} \approx \frac{I_{tail}}{2\eta N q_{max}} \tag{5.26}$$

Surprisingly, the noise of the tail current source in the vicinity of f_0 does not affect the phase noise. Rather, as will be discussed in CHAPTER 6, its low frequency noise as well as its noise in the vicinity of *even* multiples of the oscillation frequency affect the phase noise. Tail noise in the vicinity of even harmonics can be significantly reduced by a variety of means, such as with a series inductor or a parallel capacitor. As before, the effect of low frequency noise can be minimized by exploiting symmetry. Therefore, only the noise of the differential transistors and the load is taken into account, as shown in Figure 5.8. The total current noise on each single-ended node is given by

$$\frac{\overline{i_n^2}}{\Delta f} = \left(\frac{\overline{i_n^2}}{\Delta f}\right)_N + \left(\frac{\overline{i_n^2}}{\Delta f}\right)_{Load} = 4kTI_{tail}\left(\frac{1}{V_{char}} + \frac{1}{R_L I_{tail}}\right) \tag{5.27}$$

where R_L is the load resistor, $V_{char} = (V_{GS} - V_T)/\gamma$ for a balanced stage in the long channel limit and $V_{char} = E_c L/\gamma$ in the short channel regime. Assuming zero correlation among the various noise sources, the phase noise and jitter due to all $2N$ noise sources is $2N$ times the value given by (4.16) and (4.23). Using (5.4), the expression for the phase noise of the differential MOS ring oscillator is[1]

$$\mathcal{L}_{min}\{\Delta\omega\} \approx \frac{8}{3\eta} \cdot N \cdot \frac{kT}{P} \cdot \left(\frac{V_{DD}}{V_{char}} + \frac{V_{DD}}{R_L I_{tail}}\right) \cdot \frac{\omega_0^2}{\Delta\omega^2} \tag{5.28}$$

and κ is given by

$$\kappa_{min} = \sqrt{\frac{8}{3\eta}} \cdot \sqrt{N \cdot \frac{kT}{P} \cdot \left(\frac{V_{DD}}{V_{char}} + \frac{V_{DD}}{R_L I_{tail}}\right)} \tag{5.29}$$

Note that as a special case, for long channel devices, (5.28) reduces to (3.8) as predicted in [46]. The foregoing equations are valid in both long and short-channel regimes of operation with the right choice of V_{char}

Note that in contrast with the single-ended ring oscillator, a differential oscillator does exhibit a phase noise and jitter dependency on the number of stages, with the phase noise degrading as the number of stages increases for a given frequency and power dissipation. This result may be understood as a consequence of the necessary reduction in charge swing that is required to accommodate a constant frequency of oscillation at a fixed power level as N increases. At the same time, increasing the number of stages at a fixed total power dissipation demands a proportional reduction of tail current sources, which will reduce the swing, and hence q_{max}, by a factor of $1/N^2$.

5.2.3 Bipolar Differential Ring Oscillator

A similar approach allows us to derive the corresponding results for a bipolar differential ring oscillator. In this case, the power dissipation is given by (5.25) and the oscillation frequency by (5.26). The total noise current is given by the sum of collector shot noise and the load resistor noise:

$$\frac{\overline{i_n^2}}{\Delta f} = 2q_e I_C + \frac{4kT}{R} = 4kTI_{tail}\left(\frac{1}{V_{char}} + \frac{1}{R_L I_{tail}}\right) \tag{5.30}$$

1. It is also assumed that the ISF for the differential pair transistor noise sources and the ISF for the load resistors noise are the same. This equality is not generally true but is a good approximation here that leads to simple expressions.

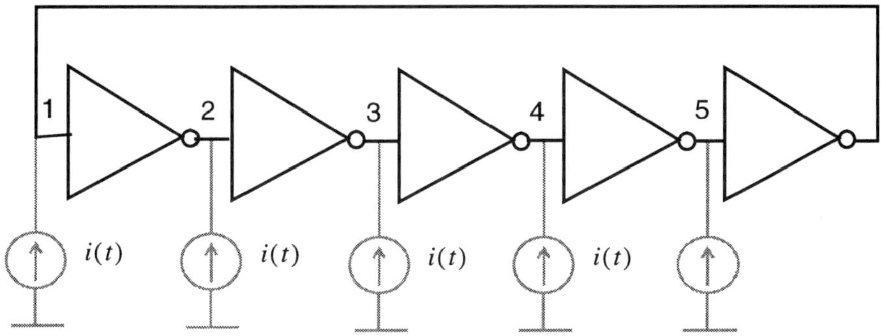

FIGURE 5.9 Five-stage ring oscillator with identical noise sources on all nodes.

where q_e is the electron charge, $I_C = I_{tail}/2$ is the collector current during the transition, and $V_{char} = 4kT/q_e$. Using these relations, the phase noise and jitter of a bipolar ring oscillator are again given by (5.28) and (5.29) with the appropriate choice of V_{char}.

5.3 Correlated Noise Sources

Noise sources on different nodes of an oscillator may be strongly correlated. This correlation can be due to various reasons. Two examples of sources with strong correlation are substrate and supply noise. These noise sources usually arise from current switching in other parts of the chip. The current fluctuations induce voltage fluctuations across the series resistance and inductance of the bondwires and pins. These fluctuations on the supply and substrate will induce a similar perturbation on different stages of the ring oscillator.

To understand the effect of this correlation, consider the special case of having identical noise sources on all the nodes of the ring oscillator as shown in Figure 5.9. If all the inverters in the oscillator are the same, the ISF for different nodes will differ only in phase by multiples of $2\pi/N$, as shown in Figure 5.10 for N=5. Therefore, the total phase due to all the sources is given by (4.4) through superposition:

$$\phi(t) = \frac{1}{q_{max}} \sum_{n=0}^{N-1} \int_{-\infty}^{t} i(\tau)\Gamma\left(\omega_0\tau + \frac{2\pi n}{N}\right) d\tau = \frac{1}{q_{max}} \int_{-\infty}^{t} i(\tau)\left[\sum_{n=0}^{N-1} \Gamma\left(\omega_0\tau + \frac{2\pi n}{N}\right)\right] d\tau \quad \textbf{(5.31)}$$

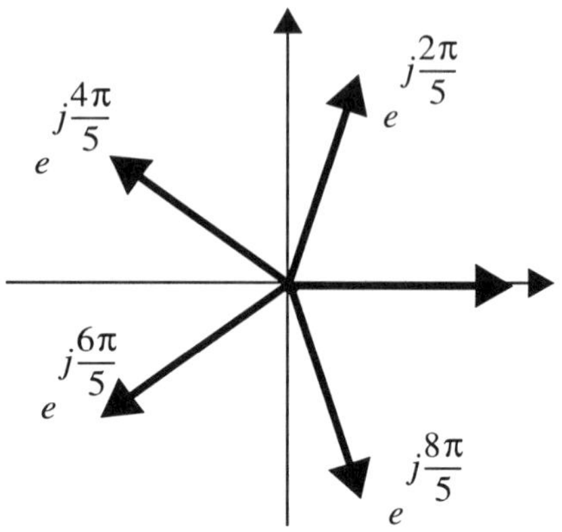

FIGURE 5.10 Phasors for noise contributions from each source.

Expanding the term in brackets in a Fourier series, it can be observed that it is zero except at dc and multiples of $N\omega_0$, *i.e.*,

$$\phi(t) = \frac{N}{q_{max}} \sum_{n=0}^{\infty} c_{(nN)} \int_{-\infty}^{t} i(\tau)\cos(nN\omega_0\tau)d\tau \qquad (5.32)$$

which means that for fully correlated sources, only noise in the vicinity of integer multiples of $N\omega_0$ affects the phase.

To verify this effect, a set of simulations similar to those in Section 4.2.2 are run. Sinusoidal currents with an amplitude of 10µA were injected into all five nodes of the ring oscillator of Figure 4.6 at different offsets from integer multiples of the frequency of oscillation and the induced sidebands were measured. The measured sideband power with respect to the carrier is plotted in Figure 5.11.

As can be seen in Figure 5.11, only injection at low frequency and in the vicinity of the 5th harmonic are integrated and show a -20dB/dec slope. The effect of injection in the vicinity of harmonics which are not integer multiples of N is much smaller than at the integer ones. Ideally there should be no sideband induced by the injection in the vicinity of harmonics that are not integer multiples of N; however, as can be seen in Figure 5.11, there is some sideband power due to the amplitude response discussed in Section 4.6.

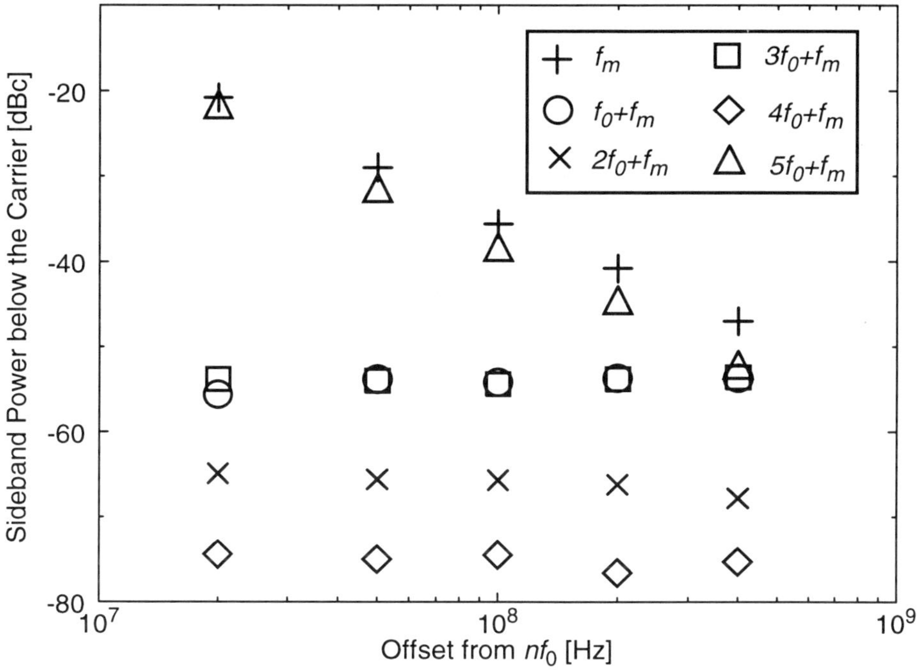

FIGURE 5.11 Sideband power below carrier for fully correlated injection at nf_0+f_m.

To demonstrate the generality of this property for other ring oscillator topologies, the more practical case of current-starved inverted-chain ring oscillators is considered. A similar simulation is performed on a 5 stage ring oscillator made of the stages shown in Figure 5.12. The effect of a sinusoidal current source on nodes *N*, *P* and *MID* at f_m and f_0+f_m is plotted vs. f_m in Figure 5.13 Again, the injection at low frequency is integrated and shows $1/f^2$ behavior, while injection in the vicinity of the oscillation frequency does not.

Low frequency noise can also result in correlation between uncertainties introduced during different cycles, as its value does not change significantly over a small number of periods. Therefore the uncertainties add up in amplitude rather than power resulting in a region with a slope of 1 in the *log-log* plot of jitter even in the absence of external noise sources such as substrate and supply noise.

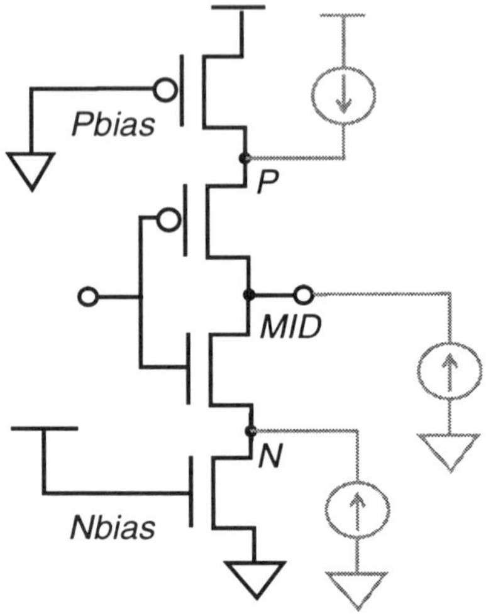

FIGURE 5.12 One stage of a current starved inverter ring oscillator.

5.4 Design Implications in Ring Oscillators

In this section, design implications for ring oscillators not mentioned in the previous sections will be discussed.

5.4.1 Differential vs. Single-Ended

A commonly asked question is the preferred topology for MOS ring oscillators, *i.e.*, which one of the single-ended or the differential topologies results in better jitter and phase noise performance for a given center frequency, f_0, and total power dissipation, P.

Equations (5.17) and (5.28) can be used to compare the phase noise performance of single-ended and differential MOS ring oscillators in the thermal noise limited case. As can be seen for N stages, the phase noise of a properly designed differential ring oscillator is approximately $N[1 + V_{char}/(R_L I_{tail})]$ times larger than the phase noise of a single-ended oscillator of equal N, P and f_0. Since the minimum N for a regular ring oscillator is three, even a properly designed differential CMOS ring oscillator

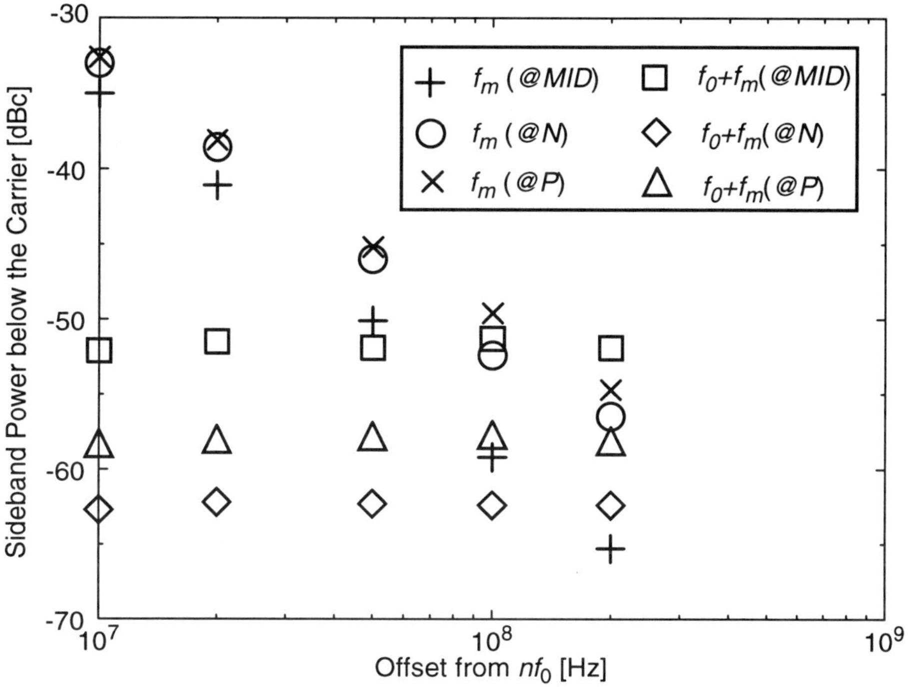

FIGURE 5.13 Sideband power below carrier for fully correlated injection into nodes, *MID*, *N* and *P* at f_m and f_0+f_m.

underperforms its single-ended counterpart, with disparity increasing with the large number of stages. This difference is even more pronounced if proper precautions to reduce the noise of the tail current are not taken.

The difference in the behavior of these two types of oscillators with respect to the number of stages can be traced to the way they dissipate power. The dc current drawn from the supply is independent of the number and slope of the transitions in differential ring oscillators. On the contrary, inverter-chain ring oscillators dissipate power mainly on a *per transition* basis and therefore have a better phase noise for a given power dissipation. This difference becomes even larger as the number of stages increases.However, a differential topology may still be preferred in ICs with a large amount of digital circuitry because of the lower sensitivity to substrate and supply noise, as well as lower noise injection into other circuits on the same chip. The decision of which architecture to use should be based on both of these considerations.

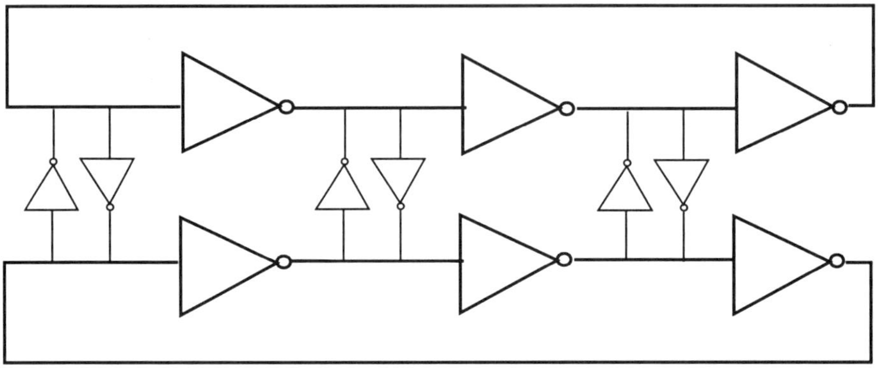

FIGURE 5.14 Coupled single-ended ring oscillators.

The common-mode sensitivity problem in a single-ended ring oscillator can be mitigated to some extent using two identical ring oscillators laid out close to each other that oscillate out of phase because of small coupling inverters [109] as shown in Figure 5.14. Single-ended configurations may be used in a less noisy environment to achieve better phase noise performance for a given power dissipation.

5.4.2 Optimum Number of Stages

Another commonly debated question concerns the optimum number of inverter stages in a ring oscillator to achieve the best jitter and phase noise for a given f_0 and P. As seen in (5.17), for single-ended oscillators, the phase noise and jitter in $1/f^2$ region are not strong functions of the number of stages for single-ended CMOS ring oscillators. However, if the symmetry criteria are not well satisfied, and/or the process has large $1/f$ noise, (5.10) predicts that a larger N will reduce the jitter. In general, the choice of the number of stages must be made on the basis of several design criteria, such as $1/f$ noise effect, the desired maximum frequency of oscillation, and the influence of external noise sources, such as supply and substrate noise, that may not scale with N.

The jitter and phase noise behavior is different for differential ring oscillators. As (5.28) suggests, jitter and phase noise increase with an increasing number of stages. Hence, if the $1/f$ noise corner is not large, and/or proper symmetry measures have been taken, the minimum number of stages (3 or 4) should be used to give the best performance. This recommendation holds even if the power dissipation is not a primary issue. It is not fair to argue that burning more power in a larger number of stages allows the achievement of better phase noise, since dissipating the same total power in a smaller number of stages with larger devices results in better jitter and phase noise, as long as it is possible to maximize the total charge swing.

5.4.3 Lowering the Effect of Correlated Noise Sources

Substrate and supply noise are among other important sources of noise and can be dominant in large digital environments. There are two major differences between these noise sources and internal device noise. First, the power spectral density of these sources is usually non-white and often demonstrates strong peaks at various frequencies[135]. Even more important is that the substrate and supply noise on different nodes of the ring oscillator have a very strong correlation. This property changes the response of the oscillator to these sources as discussed in Section 5.3.

A very important insight can be obtained from (5.32). It shows that for the correlated part of the noise, only the c_n values associated with integer multiples of number of stages, N, contribute to total phase fluctuations. Therefore, every effort should be made to *maximize* the correlated part of substrate and supply noise. This can be done by making the inverter stages and the noise sources on each node as similar to each other as possible by proper layout and circuit design.

The layout should be kept symmetrical. The inverter stages should be laid out close to each other so that substrate noise appears as a common-mode source. This consideration is particularly important in the case of a lightly doped substrate, since such a substrate may not act as a single node [135]. It is also important that the orientation of all the stages be kept identical. The interconnecting wires between the stages must be identical in length and shape.

The circuit should be designed so that the same supply line goes to all the inverter stages. Also the loading from the stages being driven should be kept identical for all the nodes, for example by using dummy buffer stages on all the nodes. A larger number of stages will also be helpful because a smaller number of c_n coefficients will affect the phase noise. Finally, the low frequency portion of the substrate and supply noise plays an important role in the jitter and phase noise. Fortunately the effect of low frequency noise can be reduced by exploiting symmetry to minimize c_0.

5.4.4 The Effect of Tail Current Noise Source

The tail current source noise in a differential structure may play an important role in the jitter and phase noise of ring oscillators. The low frequency noise of the tail current source affects phase noise if the symmetry criteria mentioned in CHAPTER 4 are not met by each half circuit. In such cases, the ISF for the tail current source has a large dc value, which increases the upconversion of low frequency noise to phase noise. This upconversion is particularly prominent if the tail device has a large $1/f$ noise corner.

As will be discussed in Section 6.3, the noise of the tail current source in the vicinity of f_0 does not affect the jitter and phase noise of the oscillator, unlike the noise of the differential pair. However, noise in the vicinity of even multiples of f_0 does affect the phase noise, implying that low pass filtering the tail current source with a series inductor can improve the phase noise of a differential ring oscillator. The price paid for this improvement is the area consumed by the inductor.

5.5 Experimental Results

The phase noise measurements in this section were performed using three different systems, an HP 8563E spectrum analyzer with phase noise measurement capability, an RDL NTS-1000A phase noise measurement system, and an HP 5500 phase noise measurement system. The jitter measurements were performed using a Tektronix CSA 803A communication signal analyzer. Table 5.1, Table 5.2, and Table 5.3 summarize the phase noise measurements. All the reported phase noise values are at a 1MHz offset from the carrier, chosen to achieve the largest dynamic range in the measurement.

TABLE 5.1 Inverter-chain ring oscillators.

Index	N	NMOS W/L µm/µm	PMOS W/L µm/µm	V_{DD}	I_{sup} mA	f_o	Pred. (5.17) *dBc/Hz*	Pred. (4.16) *dBc/Hz*	Meas. PN *dBc/Hz*
1	5	3/2	5/2	5.0	0.3	232MHz	-119.9	-117.7	-118.5
2	11	4/2	6/2	5.0	0.5	115MHz	-127.2	-126.4	-126.0
3	19	10/0.25	20/0.25	2.5	10	1.33GHz	-111.8	-113.0	-111.5

Table 5.1 shows the measurement results for 3 different inverter-chain ring oscillators. These oscillators are made of the CMOS inverters shown in Fig. 4.7a, with no fre-

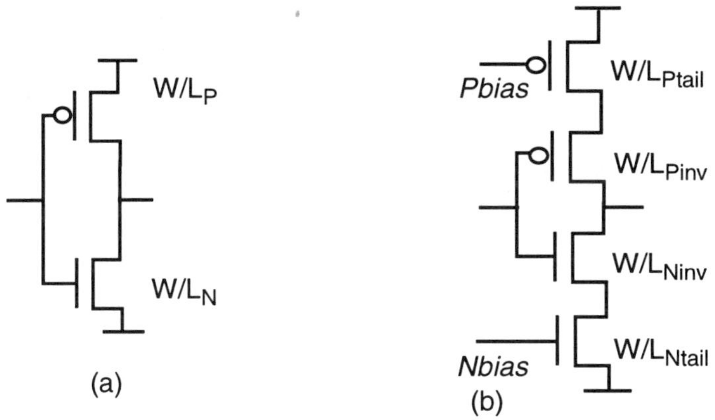

FIGURE 5.15 Stage topologies for the oscillators in Table 5.1 and Table 5.2.

quency tuning mechanism. The output is taken from one node of the ring through a few stages of tapered inverters. Oscillators number 1 and 2 are fabricated in a 2μm, 5V CMOS process, and oscillator number 3 is fabricated in a 0.25μm, 2.5Vprocess. The second column shows the number of stages in each of the oscillators. The *W/L* ratios of the NMOS and PMOS devices, as well as the supply voltages, the total measured supply currents, and the frequencies of oscillation are shown next. The phase noise prediction using (5.17) and (4.16), together with the measured phase noise, are shown in the last three columns.

As an illustrative example, the details of phase noise calculations for oscillator number 3 will be shown. Using (5.4) to calculate Γ_{rms}, the phase noise can be obtained from (4.16). The noise power is calculated when the stage is halfway through a transition. At this point, the drain current is simulated to be 3.47mA. An E_c of 4x10^6 V/m and a γ of 2.5 is used in (5.22) to obtain a noise power of $\overline{i^2}/\Delta f = 2.87 \times 10^{-22} A^2/Hz$. The total capacitance on each node is C_{total}=71.8fF and hence q_{max}=180fC. There is one such noise source on each node; therefore, the phase noise is N times the value given by (4.16), which results in $\mathcal{L}\{1MHz\} = -113.0 dBc/Hz$. The accuracy of this prediction can be further improved by direct simulation of the effective ISF and calculation of its rms value.

Table 5.2 summarizes the data obtained for current-starved ring oscillators with the cell structure shown in Figure 5.15b, all implemented in the same 0.25μm, 2.5V process. Ring oscillators with a different number of stages were designed with roughly constant oscillation frequency and total power dissipation. Frequency adjustment is achieved by changing the channel length, while total power dissipation control is per-

TABLE 5.2 Current-starved inverter-chain ring oscillators.

Index	N	*Ninv* *W/L* μm/ μm	*Pinv* *W/L* μm/ μm	*Ntail* *W/L* μm/ μm	*Ptail* *W/L* μm/ μm	I_{sup} mA	f_o MHz	Pred. (5.17) *dBc/ Hz*	Pred. (4.16) *dBc/ Hz*	Meas. PN *dBc/ Hz*
4	3	35/ 0.53	70/ 0.53	28/ 0.53	56/ 0.53	2.34	751	-113.8	-116.6	-114.0
5	5	21/ 0.39	42/ 0.39	23/ 0.39	46/ 0.39	2.51	850	-111.7	-111.9	-112.6
6	7	14/ 0.36	28/ 0.36	36.8/ 0.36	73.5/ 0.36	2.49	931	-110.5	-110.4	-111.7
7	9	12.6/ 0.32	25.2/ 0.32	28/ 0.32	56/ 0.32	2.73	932	-110.4	-113.5	-112.5
8	11	10.5/ 0.32	21/ 0.32	146/ 0.32	291/ 0.32	2.65	869	-110.9	-110.1	-112.2
9	15	9.1/ 0.28	18.2/ 0.28	146/ 0.28	291/ 0.28	2.8	929	-110.0	-110.7	-112.3
10	17	7.4/ 0.25	12.6/ 0.25	25.2/ 0.28	50.4/ 0.28	3.8	898	-111.2	-109.4	-112.0
11	19	6.3/ 0.25	12.6/ 0.25	56/ 0.25	112/ 0.25	3.9	959	-110.6	-110.1	-110.9

formed by changing device width. The *W/L* ratios of the inverter and the tail NMOS and PMOS devices are shown in Table 5.2. The node *Nbias* is kept at V_{DD}, while node *PBias* is at 0V. The measured total current dissipation and the frequency of oscillation can be found in columns 7 and 8. Phase noise calculations based on (5.17) and (4.17) are in good agreement with the measured results. The die photo of the chip containing these oscillators is shown in Figure 5.16. The slightly superior phase noise of the 3 stage ring oscillator (number 4) can be attributed to lower oscillation frequency and longer channel length (and hence smaller γ).

Table 5.3 summarizes the results obtained for differential ring oscillators of various sizes and lengths, using the inverter topology shown in Figure 5.17, and covering a large span of frequencies up to 5.5GHz. All these ring oscillators are implemented in the same 0.25μm, 2.5V process and all, except the one marked with N/A, have the tuning circuit shown in Figure 5.17. The resistors are implemented using an unsilicided polysilicon layer. The main reason for using poly resistors is to reduce $1/f$ noise upconversion by making the waveform on each node closer to the step response of an

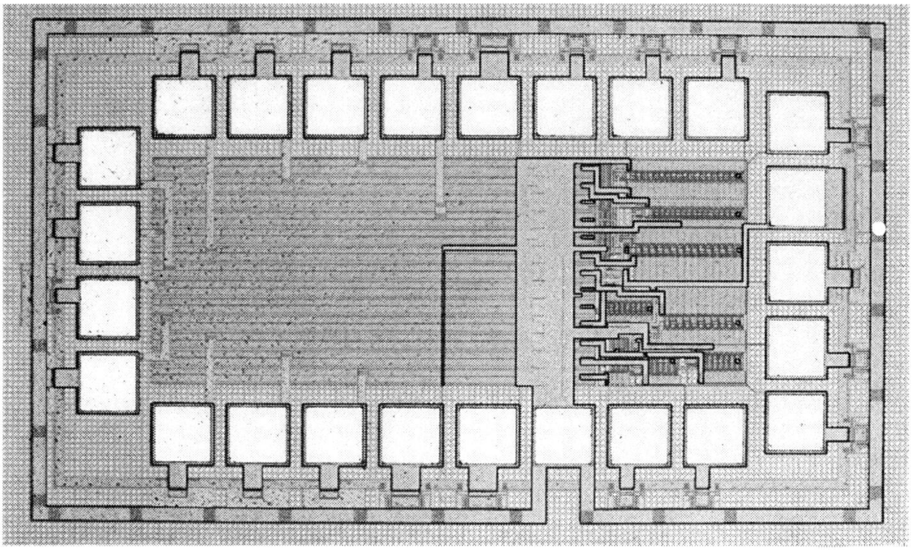

FIGURE 5.16 Die photograph of the chip containing single-ended ring oscillators.

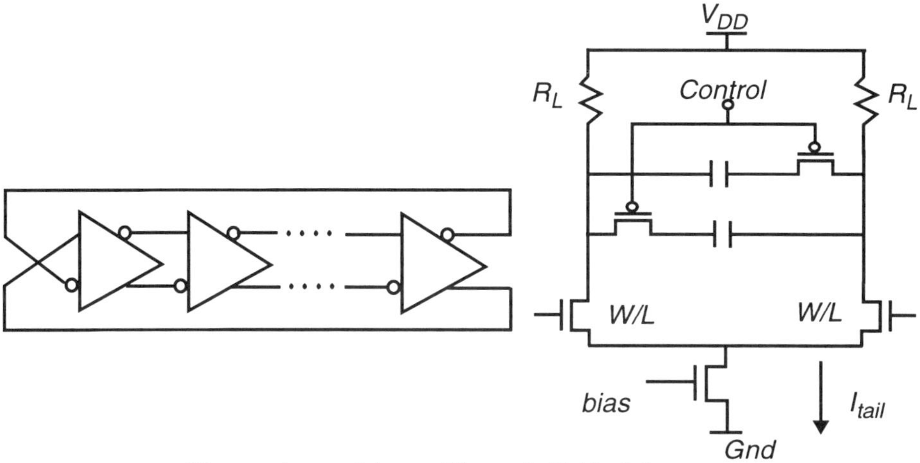

FIGURE 5.17 The topology of the oscillators in Table 5.3.

L ratios of the differential pair are shown in Table 5.3. A fixed 2.5V power supply is used, resulting in different total power dissipations. As before, the measured phase noise is in good agreement with the predicted phase noise using (5.28) and (4.16).

TABLE 5.3 Differential ring oscillators.

Index	N	W/L μm/μm	R_L Ω	I_{tail} mA	P_{tot} mW	f_{max} MHz	Tuning range	Pred. (5.28) *dBc/Hz*	Pred. (4.16) *dBc/Hz*	Meas. PN *dBc/Hz*
12	4	4.2/0.25	2k	1	10	2810	34%	-95.4	-95.5	-95.2
13	4	8.4/0.25	1k	2	20	4470	42%	-95.1	-94.0	-94.3
14	4	16.8/0.25	500	4	40	3890	44%	-98.5	-97.2	-97.4
15	4	33.6/0.25	250	8	80	5430	25%	-98.7	-99.6	-98.5
16	4	8.4/0.25	2k	1	10	2870	37%	-95.2	-96.6	-93.8
17	4	16.8/0.25	1k	2	20	3390	45%	-96.7	-97.9	-96.8
18	4	33.6/0.25	500	4	40	5330	32%	-95.8	-97.2	-95.3
19	4	16.8/0.25	2k	1	10	1750	73%	-99.5	-97.5	-95.2
20	4	33.6/0.25	1k	2	20	2240	58%	-100.3	-100.3	-99.0
21	4	33.6/0.25	2k	1	10	1270	67%	-104.4	-101.8	-100.2
22	4	67.2/0.25	1k	2	20	1190	76%	-105.8	-102.6	-100.2
23	4	33.6/0.25	2k	1	10	1530	N/A	-100.6	-98.9	-97.3
24	6	13.4/0.25	3k	0.67	10	859	58%	-103.9	-106.0	-104.3
25	8	6.7/0.25	4k	0.5	10	731	74%	-104.1	-106.3	-106.2
26	12	4.2/0.25	6k	0.33	10	447	52%	-106.6	-110.4	-109.5

Figure 5.18 compares the measured and predicted phase noise for four ring oscillators with a load resistance of 1kΩ. These correspond to rows, 13, 17, 20 and 22 in Table 5.3. The die photo of oscillator number 26 can be seen in Figure 5.19.

To illustrate further how the phase noise predictions shown in Table 5.3 are obtained, the phase noise calculations for oscillator number 12 will be shown. The noise current due to one of the NMOS transistors in the differential pair is given by (5.22). The total capacitance on each node in the balanced case is C_{total}=41.6fF and the voltage swing is 1.21V from simulation and hence, q_{max}=50.3fC. In the balanced case, this current is half of the tail current, *i.e.*, I_D=0.5mA. The noise current of the NMOS device has a single sideband spectral density of $\overline{i^2}/\Delta f = 4.14 \times 10^{-23} A^2/Hz$ and the thermal noise due to the load resistor is $\overline{i^2}/\Delta f = 8.28 \times 10^{-24} A^2/Hz$. Hence, the total current noise density is given by $\overline{i^2}/\Delta f = 4.97 \times 10^{-23} A^2/Hz$. For a differential ring oscillator with N stages, there is one such noise source on each node, therefore the

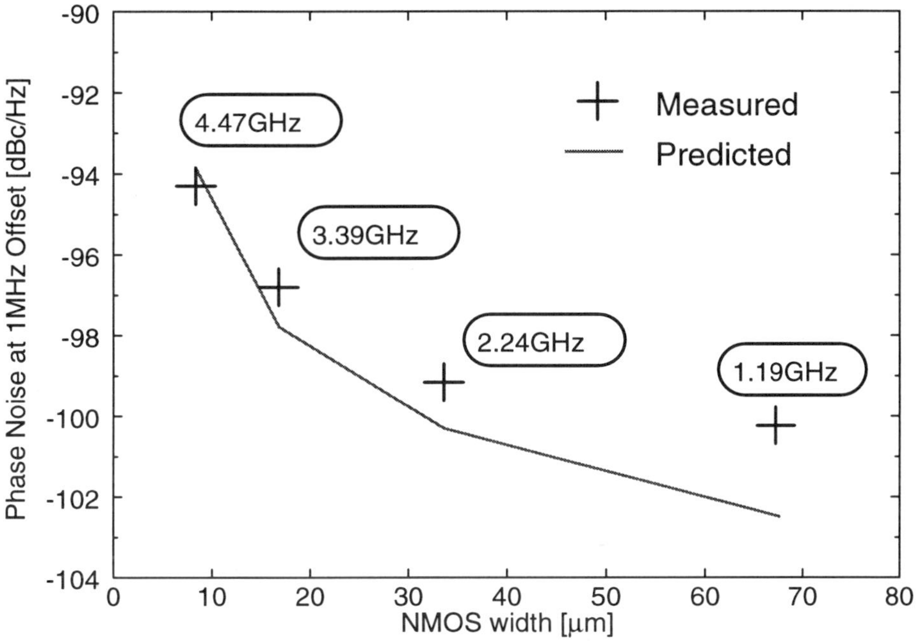

FIGURE 5.18 Predicted and measured results for single-ended rings.

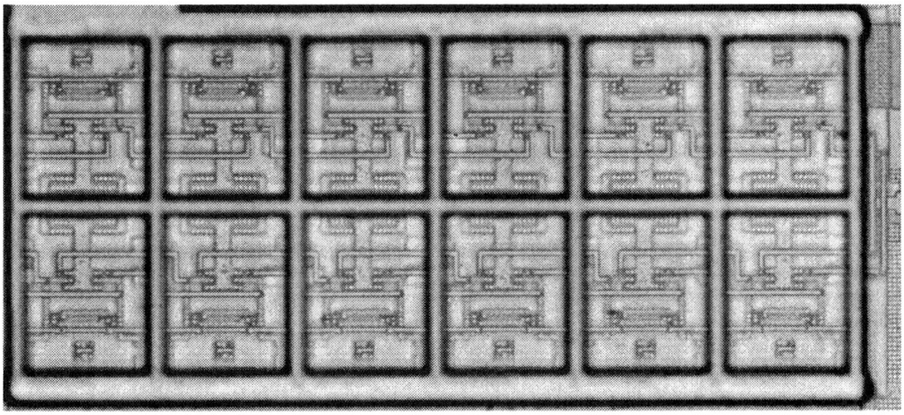

FIGURE 5.19 Die photograph of the 12 stage differential ring oscillator.

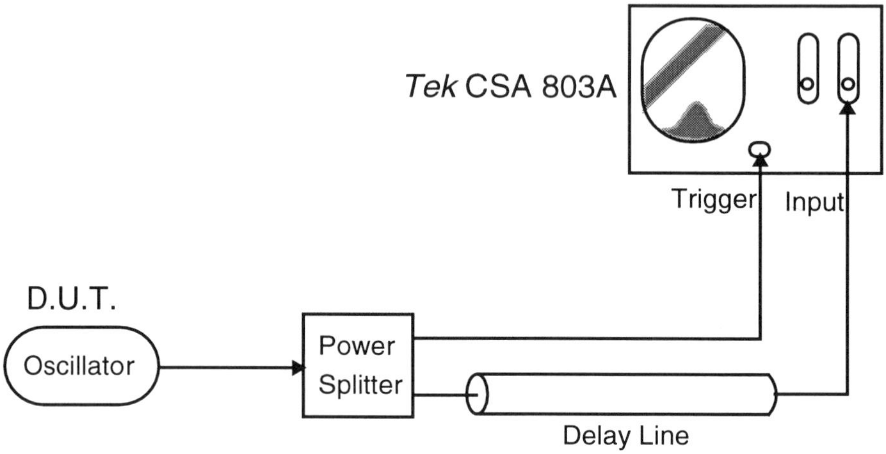

FIGURE 5.20 Timing jitter measurement setup.

power dissipation is $NV_{DD}I_{tail}$=10mW and R_L=2kΩ. Therefore, with an η of 0.9, (5.28) predicts a phase noise of $\mathcal{L}\{1MHz\} = -95.4dBc/Hz$, in good agreement with both measurement and (4.16).

Timing jitter for oscillator number 12 is measured using the setup shown in Figure 5.20. The oscillator output is divided into two equal power outputs with a power splitter. The CSA 803A is not capable of showing the edge it uses to trigger, as there is a 21ns minimum delay between the triggering transition and the first acquired sample. To be able to look at the triggering edge (and perhaps the edges before that), a delay line of approximately 25ns is inserted in the signal path in front of the sampling head. This way the very edge used to trigger the signal can be viewed. If the sampling head and power splitter were noiseless, this edge would show no jitter. However, the power splitter and the sampling head introduce noise onto the signal which cannot be easily distinguished from the DUT's jitter. This extra jitter can be directly measured by looking at the jitter on the triggering edge. This edge can be readily identified since it has lower rms jitter than the transitions before and after it. The effect of this excess jitter should be subtracted from the jitter due to the DUT. Assuming no correlation between the jitter of the DUT and the sampling head, the equivalent jitter due to the DUT can be estimated by

$$\sigma_{\tau,\,eff} = \sqrt{\sigma^2_{\tau,\,meas} - \sigma^2_{\tau,\,min}} \tag{5.33}$$

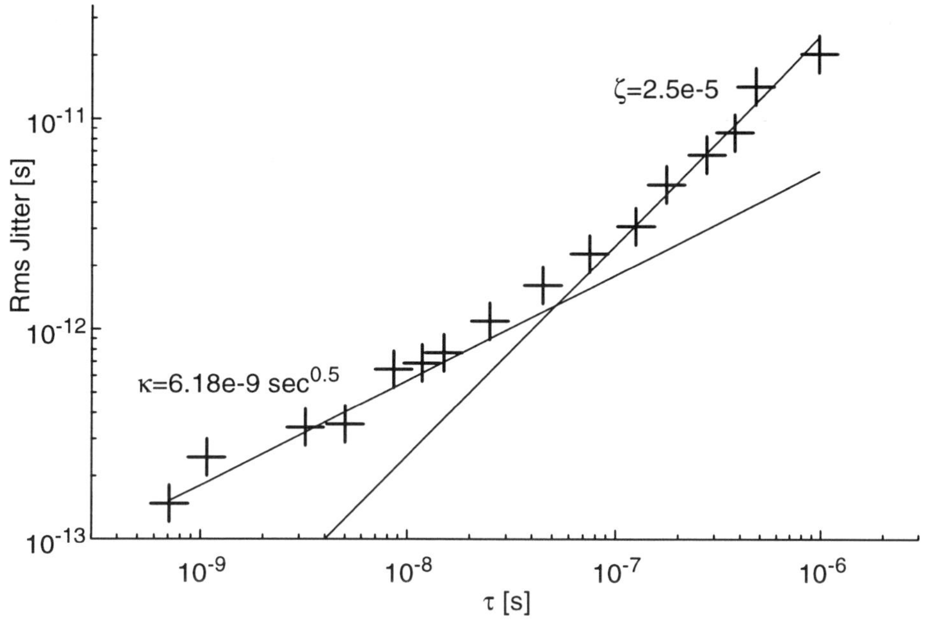

FIGURE 5.21 Timing jitter vs. delay from the trigger for oscillator number 12 (4-stage, differential ring, f_0=2.8GHz).

where $\sigma_{\tau,\,eff}$ is the effective rms timing jitter, $\sigma_{\tau,\,meas}$ is the measured rms jitter at a delay τ after the triggering edge, and $\sigma_{\tau,\,min}$ is the jitter on the triggering edge.

Figure 5.21 shows the rms jitter vs. the measurement delay for oscillator number 12 on a *log-log* plot. The best fit κ for the data shown in Figure 5.21 is $\kappa = 6.18 \times 10^{-9}\sqrt{s}$. Equations (4.23) and (5.29) result in $\kappa = 5.95 \times 10^{-9}\sqrt{s}$ and $\kappa = 6.07 \times 10^{-9}\sqrt{s}$, respectively. These results are in good agreement with the measurements.The region of the jitter plot with the slope of 1 can be attributed to the *1/f* noise of the devices, as discussed at the end of Section 5.3.

A few experiments were performed on the current-starved ring oscillators to investigate the effect of waveform symmetry on low frequency noise upconversion. The first experiment in this group investigated the effect of symmetry on $1/f^3$ region behavior. It involves a 7-stage current-starved, single-ended ring oscillator in which each inverter stage consists of an additional NMOS and PMOS device in series, as shown in Figure 5.15b. It is implemented in a 2µm, 5V CMOS process technology. The gate drives of the added transistors allow independent control of the rise and fall times. Figure 5.22 shows the phase noise, measured using a delay line based phase noise

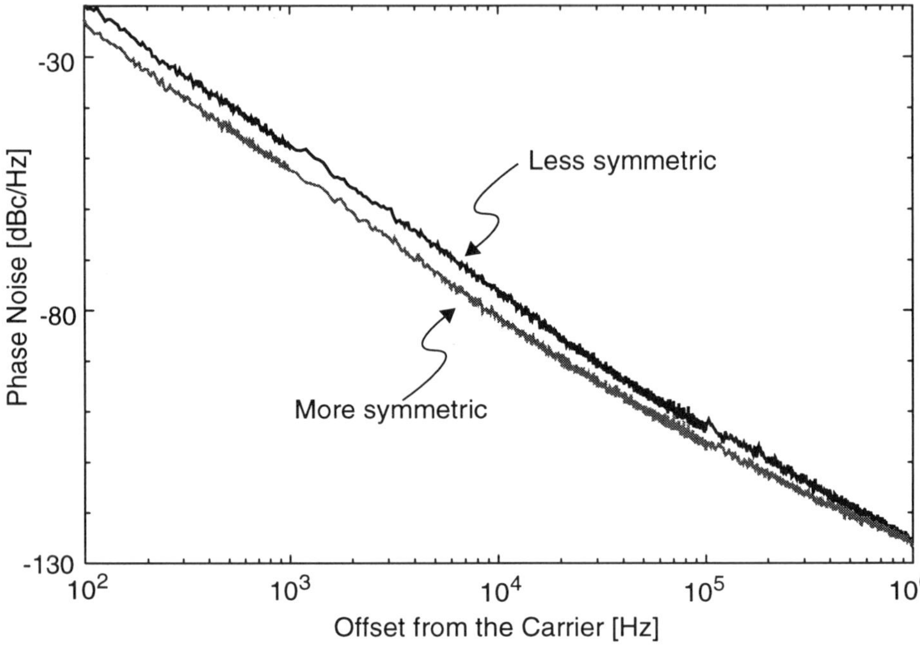

FIGURE 5.22 Effect of symmetry on the seven stage current starved ring oscillator.

measurement system [142], when the control voltages are adjusted to achieve symmetry *vs.* when they are not. In both cases the control voltages are adjusted to keep the oscillation frequency constant at 60MHz. As can be seen, making the waveform more symmetric reduces the phase noise in the $1/f^3$ region without significantly affecting the $1/f^2$ region.

Another experiment on the same circuit is shown in Figure 5.23, which shows the phase noise power spectrum at a 10kHz offset *vs.* the symmetry-controlling voltage. For all the data points, the control voltages are adjusted to keep the oscillation frequency at 50MHz. As can be seen, the phase noise reaches a minimum by adjusting the symmetry properties of the waveform. This reduction is limited by the mismatch in the transistors in different stages. Also at a given offset from the carrier, the phase noise cannot be improved beyond the noise in the $1/f^2$ region. This measurement was performed with an HP8590B spectrum analyzer.

In another experiment, a 9-stage current-starved ring oscillator with stages similar to Figure 5.15b was implemented in a 0.25μm CMOS process technology. The phase noise of the oscillator is measured for different values of *Nbias* and *PBias*. Again

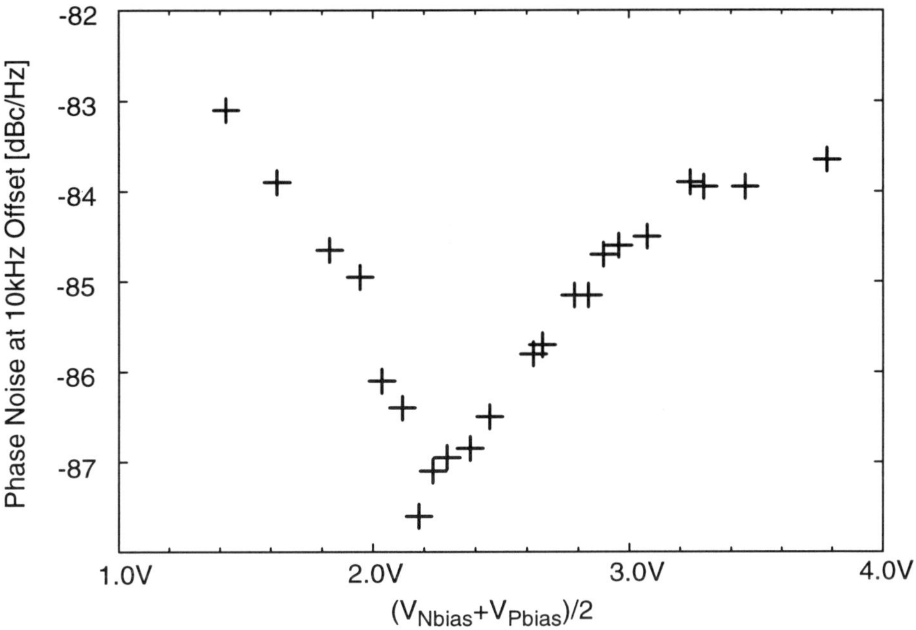

FIGURE 5.23 Phase noise sidebands at 10kHz offset vs. voltage that controls rise and fall time ratio.

these bias voltages are chosen in such a way to keep a constant oscillation frequency of 600MHz while changing only the rise-time to fall-time ratio. This time the $1/f^3$ corner of the phase noise is measured for different ratios of the pull-up and pull-down currents while keeping the frequency constant. A sharp reduction in the corner frequency at the point of symmetry can be observed in Figure 5.24.

At any given offset frequency, the achievable improvement in phase noise by adjusting the symmetry is limited by the phase noise in $1/f^2$ region. This limiting behavior can be seen in the measurement results of Figure 5.23, where a 5dB improvement is noted as the ratio of rise- to fall-time is changed. However, the $1/f^3$ corner frequency, ω_{1/f^3}, can be significantly reduced (without any fundamental limits) as can be seen from the order of magnitude reduction in corner frequency shown in Figure 5.24. In practice, this reduction is limited by process variation and device mismatch.

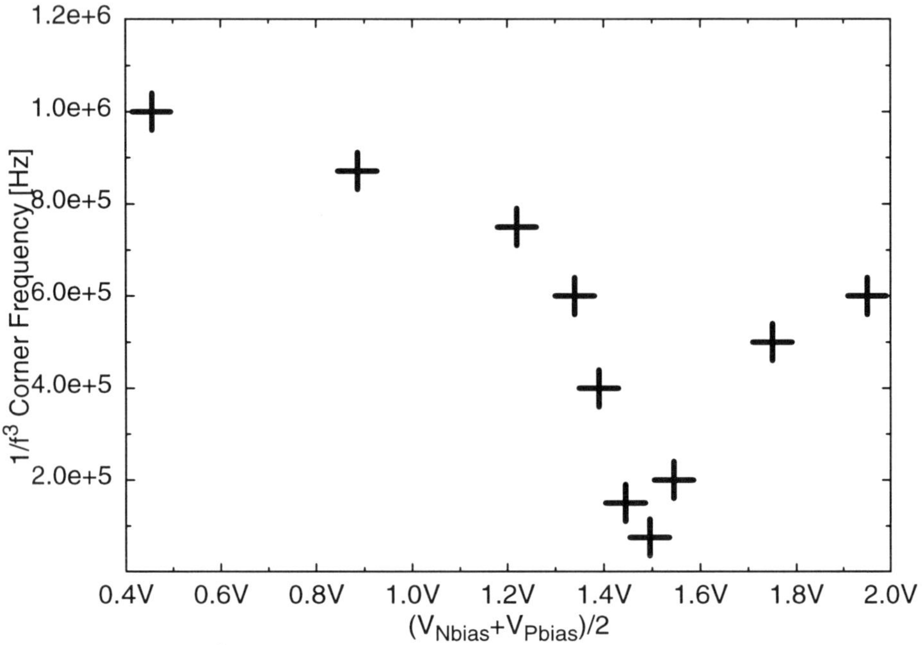

FIGURE 5.24 $1/f^3$ corner frequency vs. symmetry voltage.

5.6 Summary

Closed-form expressions for the rms and dc values of the ISF were obtained for single-ended and differential oscillators with equal and unequal rise and fall times. These approximate expressions were used to obtain approximate analytical expressions for phase noise and timing jitter in ring oscillators in terms of design parameters such as power dissipation, frequency of oscillation and number of stages. The effect of correlated noise sources on jitter and phase noise was analyzed. Design questions such as the optimum number of stages, and choice of differential vs. single-ended topology were answered. Experimental results were compared to the theoretical predictions and good agreement was observed.

CHAPTER 6 *Phase Noise in Differential LC Oscillators*

Due to their relatively good phase noise, ease of implementation and differential operation, cross-coupled LC oscillators play an important role in high frequency circuit design [51]-[56]. This chapter presents analysis and design implications in complementary cross-coupled differential oscillator [52][55]. A simple expression for the tank amplitude is obtained in Section 6.1. The effect of different noise sources in such oscillators is investigated and methods for exploiting the cyclostationary properties of noise are shown in Section 6.2. The effect of tail current noise is the subject of Section 6.3. New design implications arising from this approach and experimental results are given in Section 6.4.

6.1 Tank Amplitude

Tank voltage amplitude has an important effect on the phase noise, as emphasized by the presence of q_{max} in the denominator of (4.16). A simple expression for the tank amplitude can be obtained by assuming that the differential stage switches quickly from one side to another[1]. Figure 6.1 shows the current flowing in the complementary cross-coupled differential LC oscillator [52][55] when it is completely switched to

1. A more general describing function analysis of the oscillation amplitude can be found in Appendix F.

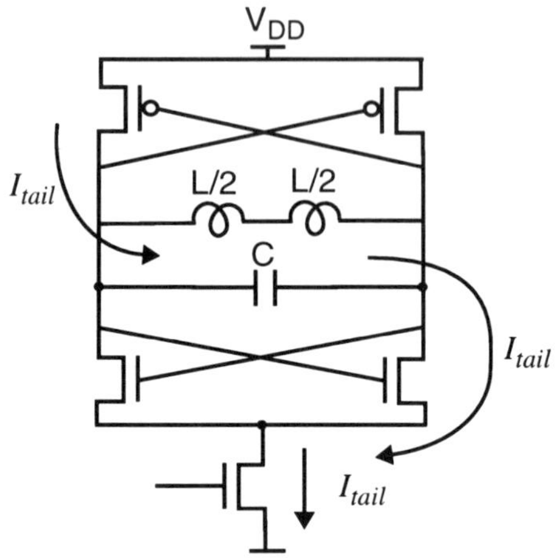

FIGURE 6.1 Differential CMOS LC oscillator.

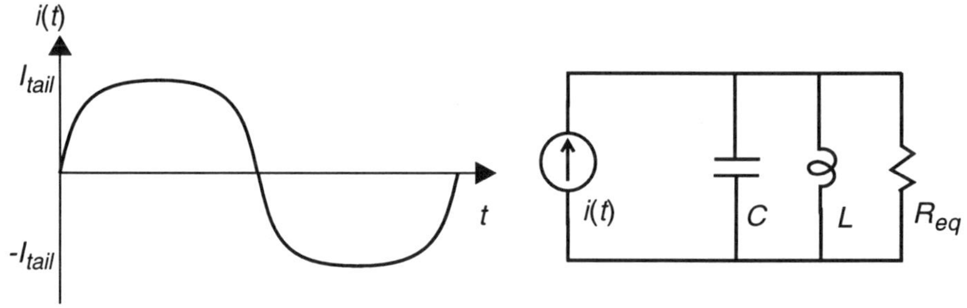

FIGURE 6.2 The equivalent circuit for the differential oscillator of Figure 6.1.

one side. As the tank voltage changes, the direction of the current flow through the tank reverses. The differential pair thus can be modeled as a current source switching between I_{tail} and $-I_{tail}$ in parallel with an RLC tank, as shown in Figure 6.2. R_{eq} is the equivalent parallel resistance of the tank.

At the frequency of resonance, the admittances of the L and C cancel, leaving R_{eq}. Harmonics of the input current are strongly attenuated by the LC tank, leaving the fundamental of the input current to induce a differential voltage swing of amplitude $(4/\pi)I_{tail}R_{eq}$ across the tank if one assumes a rectangular current waveform. At high frequencies, the current waveform may be approximated more closely by a sinusoid due to finite switching time and limited gain. In such cases, the tank amplitude can be better approximated as

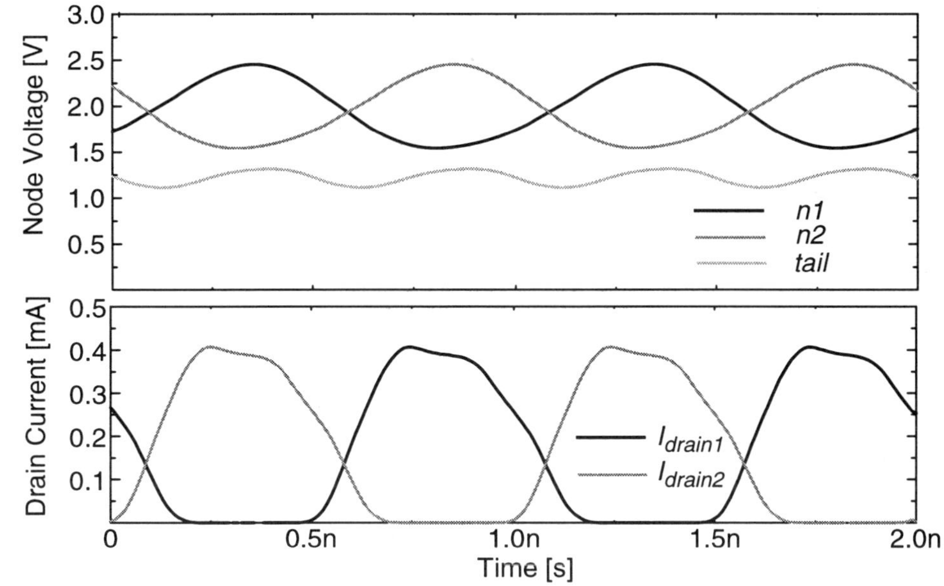

FIGURE 6.3 Simulated voltages and currents in the current limited regime.

$$V_{tank} \approx I_{tail} R_{eq} \tag{6.1}$$

This mode of operation is referred to as the "current limited" regime of operation since in this regime, the tank amplitude is solely determined by the tail current source and the tank equivalent resistance. Figure 6.3 shows the simulated node voltages as well as the drain currents of the NMOS transistors, M_1 and M_2, in this regime of operation. The values of L and C are such that the circuit oscillates at 1GHz.

Note that (6.1) loses its validity as the amplitude approaches the supply voltage because both NMOS and PMOS pairs will enter the triode region at the peaks of the voltage. Also the tail NMOS transistor may spend most (or even all) of its time in the linear region. This behavior can be seen in the simulated voltages and currents shown in Figure 6.4. The tank voltage will be clipped at V_{DD} by the PMOS transistors and at ground by the NMOS transistors. Therefore, for the oscillator of Figure 6.1, the tank voltage amplitude does not significantly exceed V_{DD}. Note that since the tail transistor is in the triode region the current through the differential NMOS transistors can drop significantly when their drain-source voltage becomes very small, as shown in Figure 6.4. This region of operation is therefore known as the "voltage limited" regime.

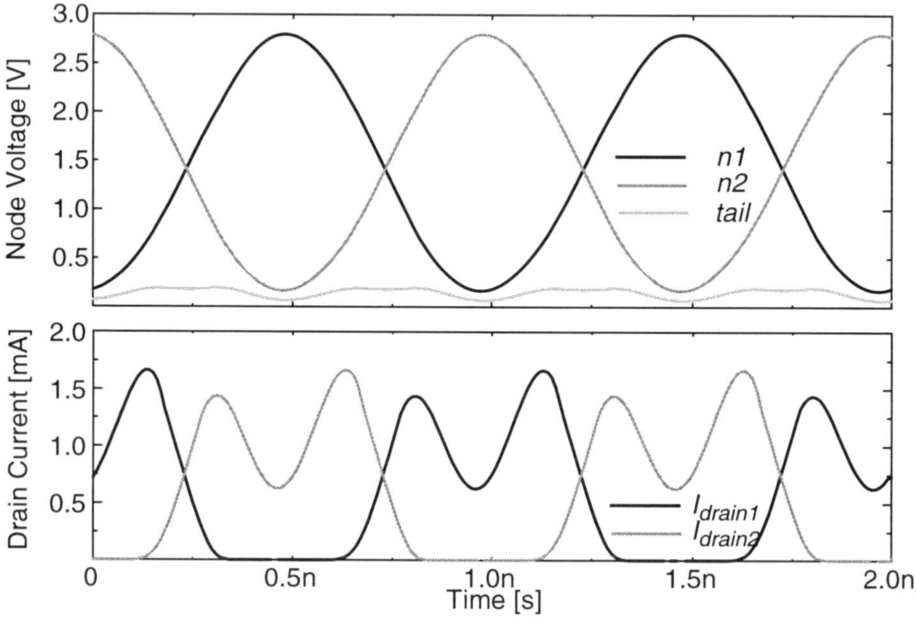

FIGURE 6.4 Simulated voltages and currents in the voltage limited regime.

Figure 6.5 shows the simulated tank voltage amplitude as a function of tail current for three different values of V_{DD}. As can be seen, the tank amplitude is proportional to the tail current in the current limited region, while it is limited by V_{DD} in the voltage limited regime.

6.2 Noise Sources

Figure 6.6 depicts the noise sources in the oscillator. The noise power densities for these sources are required to calculate the phase noise using (4.16). In general, these noise sources are cyclostationary because of the periodic changes in currents and voltages of the active devices. In this section, a simplified stationary model for the noise sources is introduced first and then subtleties arising from their true cyclostationary behavior are examined.

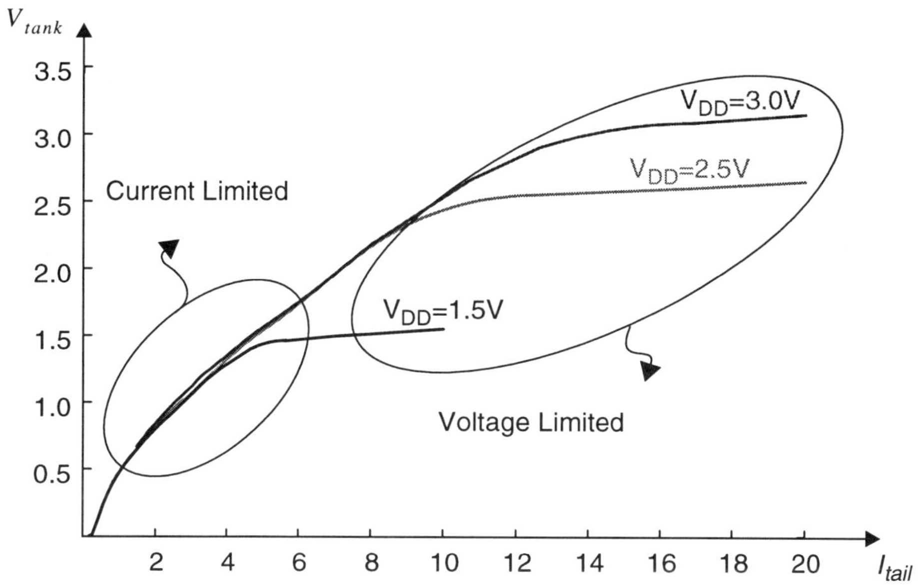

FIGURE 6.5 Simulated tank voltage versus tail current.

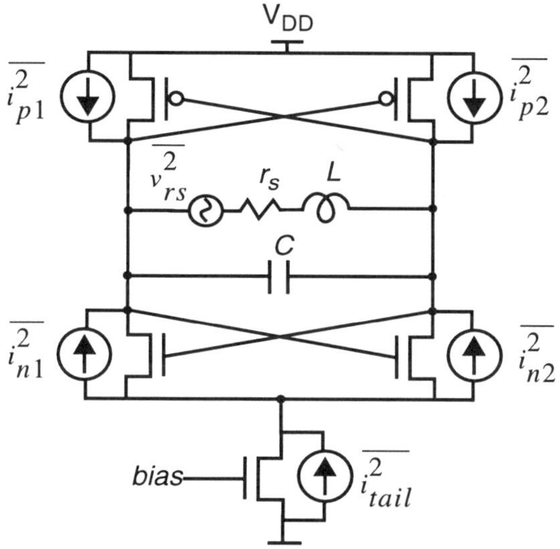

FIGURE 6.6 The noise sources in the differential CMOS LC oscillator.

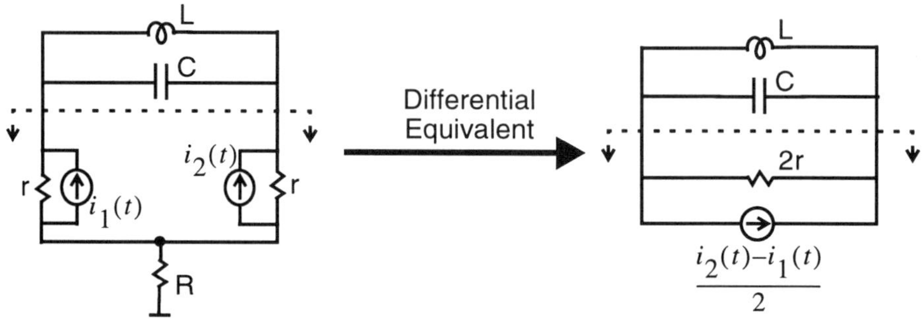

FIGURE 6.7 a) Simplified model for transistor noise sources, b) Differential equivalent circuit.

6.2.1 Stationary Noise Sources

In a simplified stationary approach, the power densities of the noise sources can be evaluated at the most sensitive time (*i.e.*, the zero crossing of the differential tank voltage) to estimate the effect of these sources. Figure 6.7a shows a simplified model of the sources in this balanced case. Converting the current sources to their Thévenin equivalent and writing KVL one obtains the equivalent differential circuit shown in Figure 6.7b. Note that the parallel resistance is canceled by the negative resistance provided by the positive feedback. Therefore the total differential noise power due to the four cross-coupled transistors is

$$\overline{i_{cc}^2} = \frac{1}{4}(\overline{i_{n1}^2} + \overline{i_{n2}^2} + \overline{i_{p1}^2} + \overline{i_{p2}^2}) = \frac{1}{2}(\overline{i_n^2} + \overline{i_p^2}) \tag{6.2}$$

where $\overline{i_n^2} = \overline{i_{n1}^2} = \overline{i_{n2}^2}$ and $\overline{i_p^2} = \overline{i_{p1}^2} = \overline{i_{p2}^2}$. Noise densities $\overline{i_n^2}/\Delta f$ and $\overline{i_p^2}/\Delta f$ are given by

$$\overline{i_n^2}/\Delta f = 4kT\gamma\mu C_{ox}\frac{W}{L}(V_{GS} - V_T) \tag{6.3}$$

where μ is the mobility of the carriers in the channel, C_{ox} is the oxide capacitance per unit area, W and L are the width and length of the MOS transistor, respectively, V_{GS} is the dc gate-source voltage and V_T is the threshold voltage. Equation (6.3) is valid for both short and long channel regimes of operation. However γ is around 2/3 for long channel transistors while it may be between 2 and 3 in the short channel region [134].

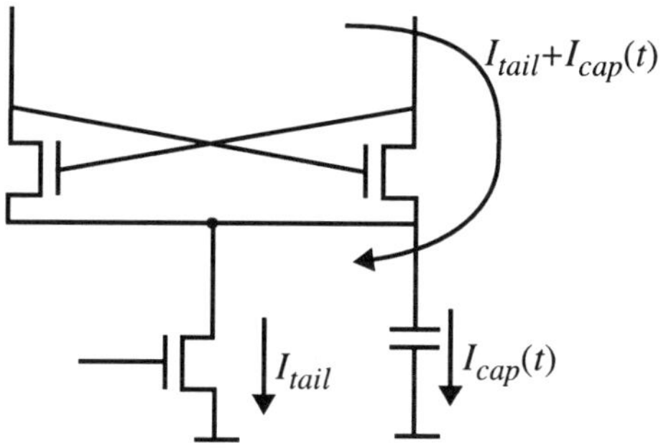

FIGURE 6.8 The effect of tail capacitor.

In addition to these sources, the contribution of the effective series resistance of the inductor, r_s, caused by ohmic losses in the metal and substrate is given by

$$\overline{i_{rs}^2}/\Delta f = 4kT\frac{r_s C}{L} = \frac{4kT}{R_p} \tag{6.4}$$

where $R_p \approx Q^2 r_s = (L\omega_0)^2/r_s$ is the equivalent parallel resistance at the frequency of oscillation.

6.2.2 Cyclostationary Behavior in the Presence of a Tail Capacitor

The foregoing analysis is based on the assumption that the sum of the currents through the differential transistors is equal to the tail current at all times. However, this assumption can break down if there is a capacitor in parallel with the tail current source, as shown in Figure 6.8. This capacitor provides an alternative path for the tail current. If the tail capacitor is large, the differential pair transistors might carry very little current for a fraction of the cycle.

To investigate further the effect of this capacitor, the simulation of the 1GHz complementary LC oscillator of Figure 6.3 was repeated with a 10pF tail capacitor. Figure 6.9 shows the voltage of the differential and tail nodes, as well as the drain currents of M_1 and M_2 in the presence of the tail capacitor. A reduction in the duty-cycle of the drain current waveform can be seen in Figure 6.9 relative to that in Figure 6.3. This small change in the duty-cycle of the waveform is particularly important as it reduces the drain current (and the drain current noise) of the differential NMOS and PMOS transistors during the zero-crossing of the differential tank voltage. As shown in CHAPTER 4, this moment is when the oscillator is most sensitive to a perturbation.

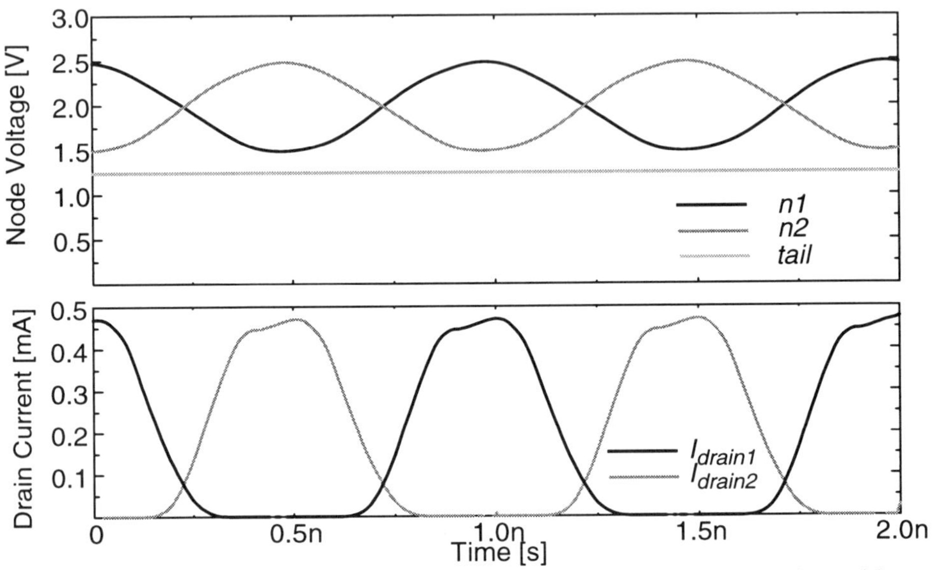

FIGURE 6.9 Simulated voltages and currents in the current limited regime with a 10pF tail capacitor.

Therefore, using an extra tail capacitor can improve the phase noise behavior of the differential LC oscillator[1]. The tail capacitor also shapes the effect of noise in M_{tail} in other important ways, as will be discussed in the next section.

In the voltage limited regime, the drain current of the differential pair transistors drops whenever the differential pair is switched to one side. This reduction in drain current reduces the noise current in the NMOS transistor. However, a reduction in noise power during the zero crossing is much more important, since it reduces the noise power during the least sensitive time, *i.e.*, the peak of the differential tank voltage. It can be verified through simulation that, in the voltage limited regime, adding a tail capacitor indeed does not reduce the duty-cycle or noise significantly.

1. The tail capacitor also attenuates the voltage variations on the *tail* node and therefore reduces the channel length modulation of the tail NMOS. It results in more symmetric waveforms and smaller harmonic distortion in the output waveform of the oscillator.

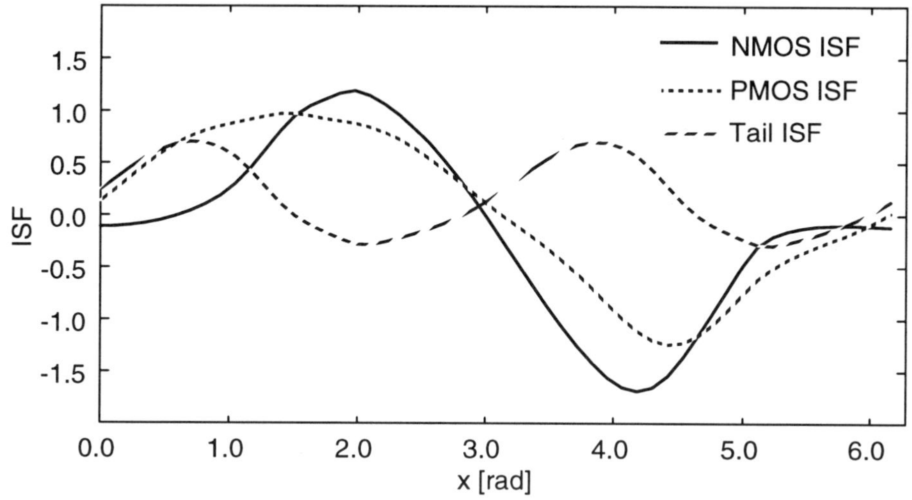

FIGURE 6.10 The simulated waveform and the ISF of various nodes and noise sources.

6.3 Tail Current Noise Source

To gain further insight into the effect of the tail noise source, its ISF as well as those for the noise sources of the NMOS and PMOS drain noise sources are shown in Figure 6.10. The ISFs are calculated using direct impulse injection and measuring the resultant phase shift.

As can be seen from Figure 6.10, the ISF associated with the tail current source has a fundamental frequency which is double the oscillation frequency. This doubling is expected since the tail node is pulled up every time each one of the differential NMOS transistors turns on and thus the tail node moves at twice the frequency of the differential voltage.

Due to this frequency doubling, the c_1 coefficient for the tail ISF is zero and therefore the noise of the tail current source in the vicinity of ω_0 has no effect on the differential noise current. However, even-order coefficients such as c_2 are significant, and therefore noise components around even harmonics of ω_0 have a significant effect on the phase noise, as shown in Figure 6.11. Also the low frequency noise component of the tail noise source can affect phase noise through asymmetry as suggested by (4.26). To verify this behavior, a sinusoidal current with an amplitude of 200μA was injected in parallel with the tail current source and the induced sideband power below the carrier

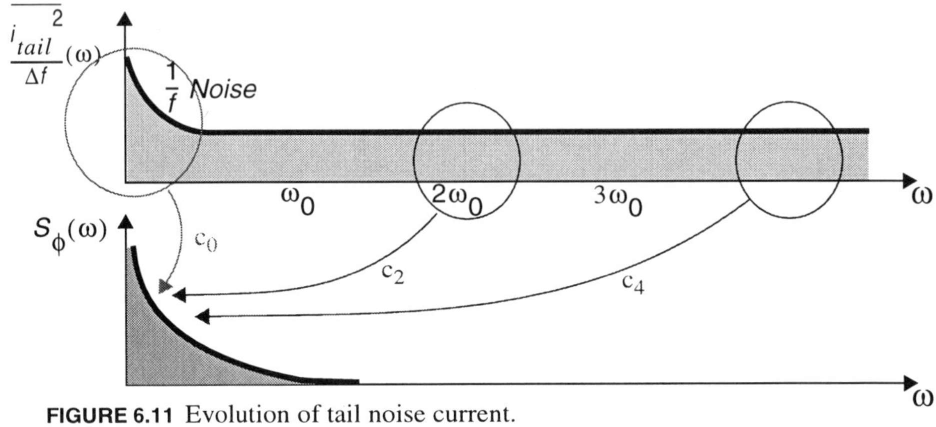

FIGURE 6.11 Evolution of tail noise current.

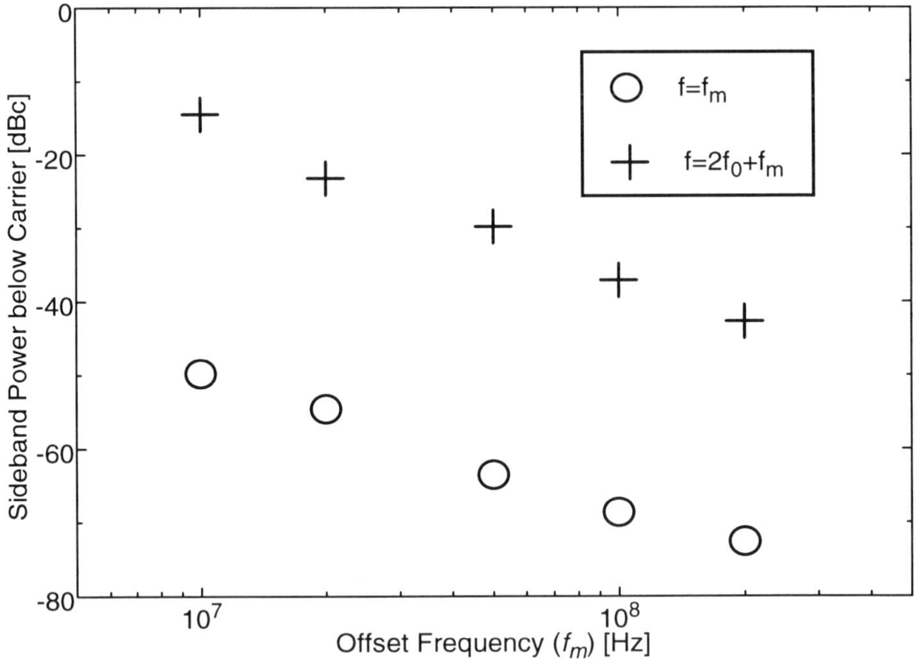

FIGURE 6.12 Induced sidebands due to sinusoidal perturbation at f_{dev} and $2f_0+f_{dev}$.

was measured using FFT analysis in HSPICE. As can be seen in Figure 6.12, sinusoidal injection at low frequency (f_m) and in the vicinity of twice the oscillation frequency ($2f_0+f_m$) results in noticeable sidebands. However, sinusoidal injection of the same amplitude at f_0+f_m does not produce any observable sidebands.

The tail capacitor mentioned in the previous section attenuates the high frequency noise components of the tail current source, and so one expects corresponding attenuation of phase noise due to this noise. In fact, the induced sidebands due to injection at $2f_0+f_m$ in the presence of the 10pF tail capacitor are below the numerical noise floor of the FFT operation.

Since upconversion of $1/f$ noise is thus the most significant remaining noise component of noisy tail current, one must properly size the tail current transistor and satisfy the single-ended symmetry criterion by sizing the cross-coupled NMOS and PMOS transistors properly.

6.4 Experimental Results and Design Implications

The complementary oscillator of Figure 6.1 was implemented in a five-metal, 0.25µm epi CMOS technology. The complementary structure is used to maintain symmetry of each half circuit to mitigate the upconversion of $1/f$ noise. Figure 6.13 shows the die photograph of the implemented oscillator. Two square inductors in series are laid out symmetrically in metal 3, 4 and 5. The series combination of the two constitutes the tank inductor. Each inductor is 230µm on a side and has four turns. Vias are used to keep the three metal layers at the same potential, and are interleaved to minimize the parasitic capacitance, as shown in Figure 6.14. Field solver simulation of this inductor predicts an inductance of 2.0nH and an effective Q of 7.5 at 1.8GHz, which translates to an effective series resistance of r_s=3.0Ω for each inductor. Simulated tank voltage amplitude vs. tail current at 1.8GHz is shown in Figure 6.5.

Figure 6.15 shows a plot of phase noise at 600kHz vs. the tail current with a 3.0V supply. The dashed line shows the phase noise predictions obtained using a simplified model for noise and amplitude, and assuming a sinusoidal waveform so that Γ^2_{rms} equals 0.5. As can be seen, these simplifying assumption lead to reasonable predictions. More accurate predictions can be obtained by calculating the true ISF and taking into account the effect of cyclostationarity of noise sources in (4.16). The solid line shows the predictions obtained using the full-blown analysis. As can be seen, very good agreement between the theoretical predictions and measurements is observed for different bias points.

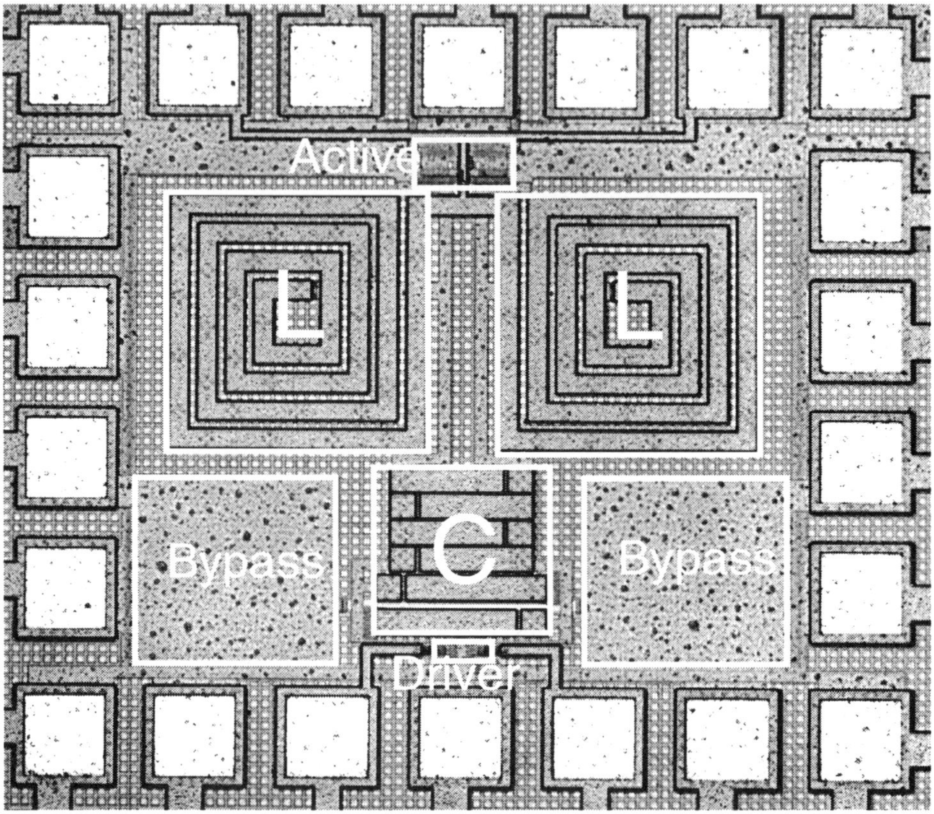

FIGURE 6.13 Die photograph of the differential LC oscillator.

To gain more insight about the trade-offs involved, the phase noise at 600kHz offset is measured for different values of the supply voltage and tail current, as shown in Figure 6.16. Each measured value is shown as a node on the three-dimensional surface. Note that bias points not achievable are shown as a flat surface. As can be seen from this graph, increasing the tail current will improve the phase noise due to the increase in oscillation amplitude. Also as can be seen, the improvement slows down as the tank voltage amplitude approaches the supply voltage. It can also be seen that the phase noise has a weak dependence on the supply voltage, improving somewhat for lower voltages. This behavior may be attributed to smaller voltage drops across the channel on the MOS transistors which reduces the effect of velocity saturation in the short channel regime and hence lowers γ.

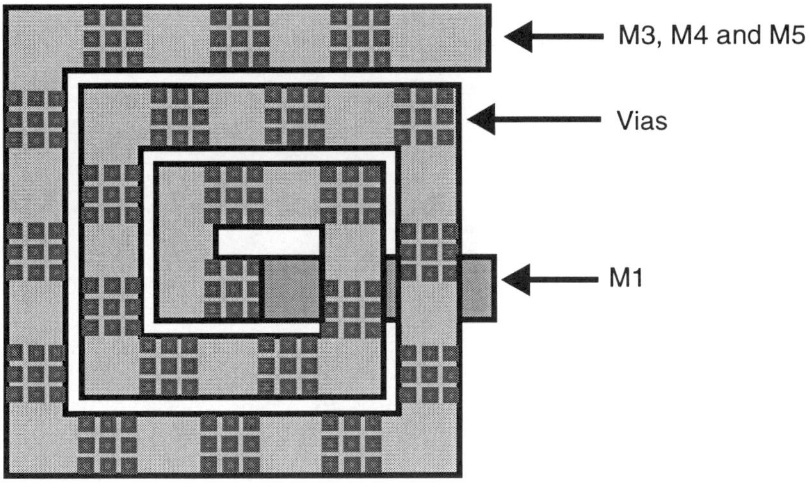

FIGURE 6.14 Via-interleaving in the spiral inductor.

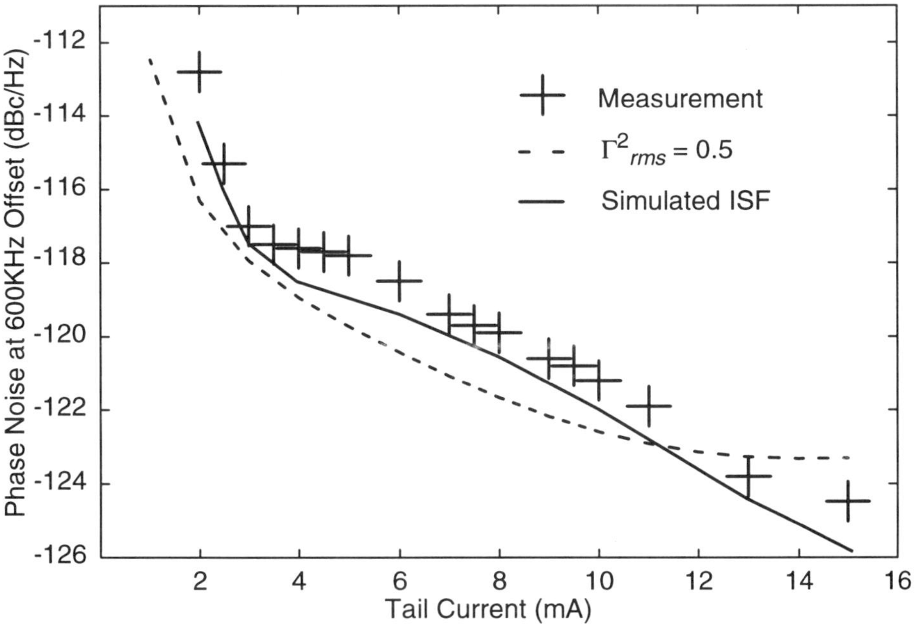

FIGURE 6.15 The phase noise at 600kHz offset for different tail currents (V_{DD}=3V).

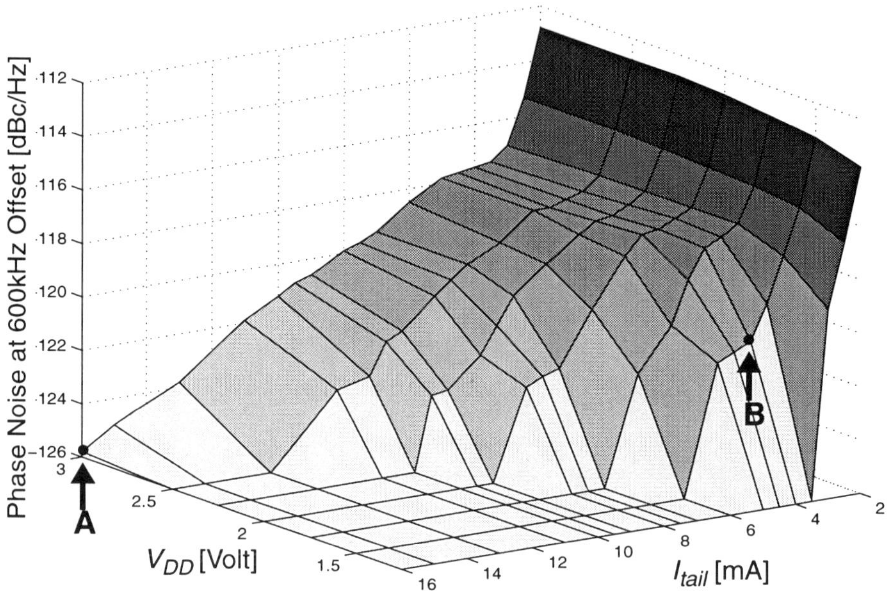

FIGURE 6.16 Measured phase noise for various supply voltage and tail current values.

The power dissipation increases as the operation point moves toward higher tail currents and supply voltage, which corresponds to moving from right to left in Figure 6.16. If the design goal is to achieve the minimum phase noise without any concern for power dissipation, the oscillator should be operated at high supply voltage and high current to allow the maximum possible tank voltage amplitude. Point A in Figure 6.16 is an example of such an operation point. It corresponds to a tail current of 16mA and a supply voltage of 3V, and results in a phase noise of -125.7dBc/Hz at 600kHz offset. However, power usually is a concern, so a more practical goal maybe to achieve the best phase noise for a given power. Equation (4.16) suggests that it is desirable to operate at the largest tank amplitude for a given power dissipation. However, the tank amplitude cannot be increased beyond V_{DD} due to voltage limiting. Therefore, according to this simple model, it is desirable to operate at the edge of the voltage limited mode of operation. As can be seen in Figure 6.5, point B in Figure 6.16 is at the verge of voltage limiting, which explains its good phase noise for a given power. Under this operation point, 4mA of dc current is drawn from a 1.5V power supply, resulting in a phase noise of -121dBc/Hz at 600kHz offset from the car-

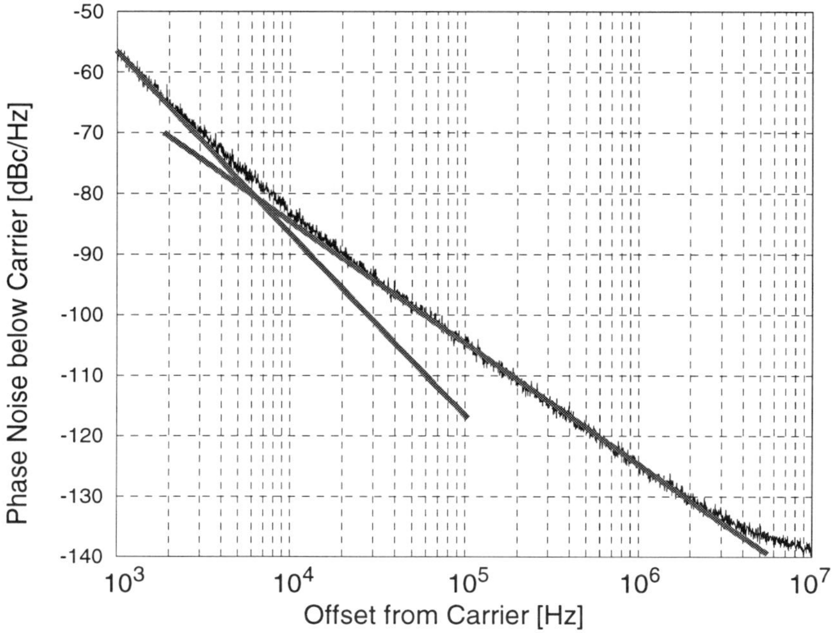

FIGURE 6.17 Measured phase noise spectrum for I_{tail}=4mA, V_{DD}=1.5V and f_0=1.8GHz.

rier, while dissipating 6mW of power. Figure 6.17 shows the phase noise sidebands vs. offset frequency for the complementary differential LC oscillator operating at point B.

To investigate the effect of the PMOS transistors, an NMOS-only oscillator is compared to the complementary case. The supply voltage is provided through the middle node of the inductor and the phase noise of this NMOS-only oscillator is measured for different supply voltages and tail currents. The result is plotted together with the data from Figure 6.16 in Figure 6.18. Note that the NMOS-only oscillator exhibits inferior phase noise for all the measured bias points.

There are several reasons for the superiority of the complementary structure over the all-NMOS structure. The complementary structure offers higher transconductance for a given current, which results in faster switching of the cross-coupled differential pair. It also offers better rise and fall time symmetry, which results in less upconversion of $1/f$ noise and other low frequency noise sources. Also the dc voltage drop across the channel is larger for the all-NMOS structure since the dc value of the drain voltage is

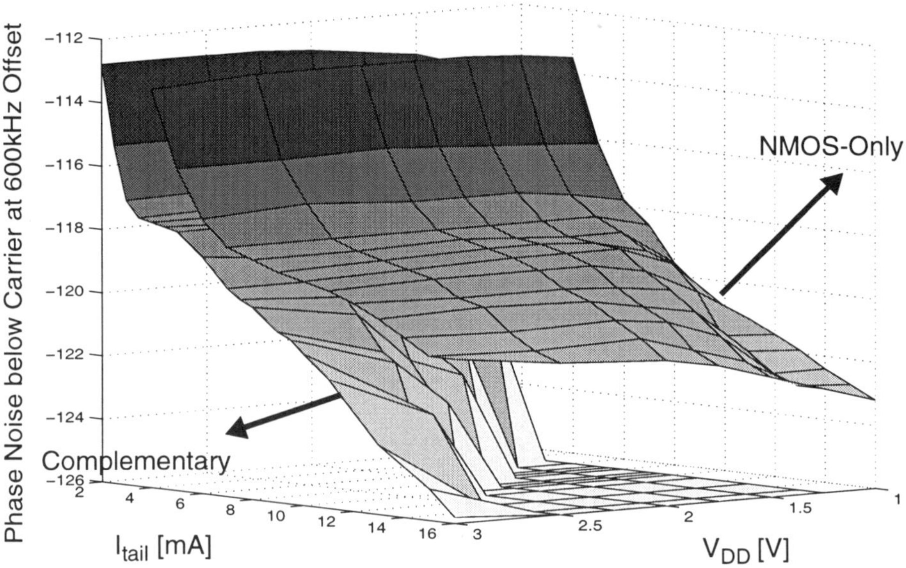

FIGURE 6.18 Measured phase noise of the complementary vs. NMOS-only topology.

V_{DD}. There is therefore stronger velocity saturation and a larger γ. As long as the oscillator operates in the current limited regime, the tank voltage swing is the same for both oscillators[1].

6.5 Summary

An analysis of phase noise in differential cross-coupled LC oscillators was presented. The effect of tail current and equivalent tank loading on voltage amplitude was shown in both current and voltage limited modes of operation. The effect of various noise

1. If the rare case of achieving the lowest possible phase noise without any concern for power dissipation is the design objective, all-NMOS structures can offer a larger voltage swing and therefore may be the preferred topology.

sources in the circuit was analyzed and it was shown that the effective noise introduced by the transistors in the differential pair can be reduced by exploiting cyclostationary properties of the sources. The predictions made are in good agreement with the measurements for different tail currents and supply voltages. Finally a 1.8GHz LC oscillator using on-chip spiral inductors exhibiting a phase noise of -121dBc/Hz at 600kHz while dissipating 6mW of power was shown.

CHAPTER 7

Extension of the Model to Multiple Noise Sources

The phase impulse response was introduced in CHAPTER 4. In this chapter the phase impulse response of an oscillator is extended to multiple inputs and is used to calculate the phase noise of an oscillator in the presence of multiple noise sources with arbitrary correlation and cyclostationarity.

7.1 Phase Response

An oscillator can be fully described by a time-dependent state-vector in an n-dimensional state-space, *i.e.*,

$$X(t) = \begin{bmatrix} x_1(t) \\ x_2(t) \\ \dots \\ x_n(t) \end{bmatrix} \tag{7.1}$$

where x_i represents the ith state variable and t is the time. For small external excitations, the dynamics of the system is governed by an equation of the following form:

$$\dot{X}(t) = f[X(t)] + b(t) \tag{7.2}$$

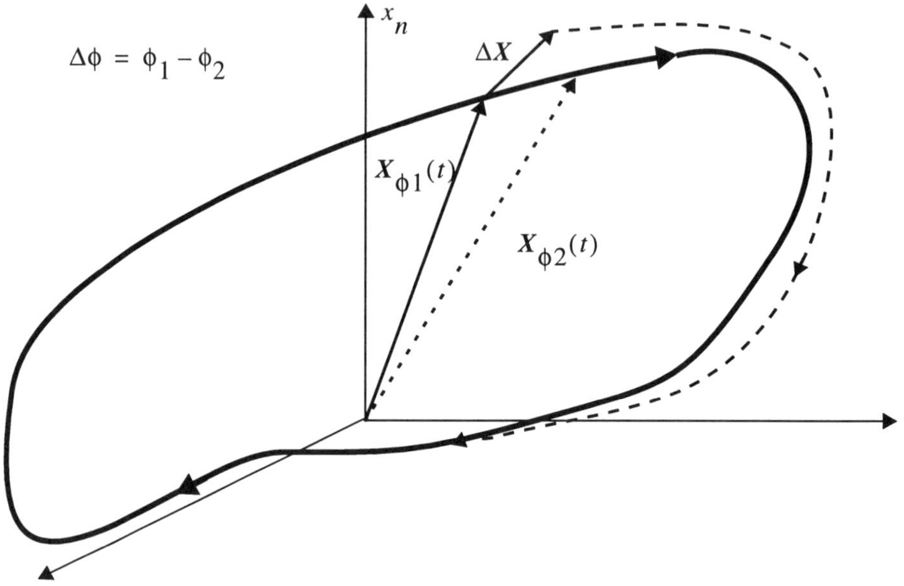

FIGURE 7.1 Limit cycle of an arbitrary oscillator in the state-space, showing the effect of an arbitrary perturbation on the phase of the oscillator.

where $\boldsymbol{f}$ is a nonlinear vector function of the state-vector and $\boldsymbol{b}$ is a vector representing the external inputs to the system. For a stable single-mode oscillator in the absence of external excitations, the state vector, $\boldsymbol{X}(t)$, periodically traverses a closed trajectory in the state-space known as the limit cycle. This steady-state situation is shown in Figure 7.1 with a solid line. The family of solutions for (7.2) are periodic, *i.e.*,

$$\boldsymbol{X}_{\phi}(t) = \boldsymbol{X}_{\phi}(t+T) \tag{7.3}$$

where T is the period of oscillation. The index ϕ identifies the phase (epoch) of a particular solution. Note that any phase-shifted version of one solution is yet another steady-state solution, *i.e.*,

$$\boldsymbol{X}_{\phi_1}(t) = \boldsymbol{X}_{\phi_2}\left(t + T\frac{\phi_1 - \phi_2}{2\pi}\right) = \boldsymbol{X}_{\phi_2}\left(t + \frac{\phi_1 - \phi_2}{\omega_0}\right) \tag{7.4}$$

where $\omega_0 = 2\pi/T$ is the steady-state angular frequency of the oscillator.

In an electrical oscillator, it is natural to choose as state-variables the voltages across the capacitors and currents through the inductors. For such a circuit, the external excitation sources can be presented as current sources in parallel with capacitors and voltage sources in series with inductors. As an illustration, imagine a state-vector in the following form:

$$X(t) = \begin{pmatrix} V_{C1}(t) \\ I_{L1}(t) \\ \dots \\ V_{Ck}(t) \end{pmatrix} \tag{7.5}$$

For the general case of voltage-dependent capacitors and inductors, the excitation vector will be given by

$$b(t) = \begin{pmatrix} i_{C1}(t) \\ v_{L1}(t) \\ \dots \\ i_{CM}(t) \end{pmatrix} \tag{7.6}$$

where $i_{Ck}(t)$ represents the perturbation current source in parallel with the time-variant capacitor C_k(t) and $v_{Lm}(t)$ represents the external perturbation voltage source in series with the time-variant inductor $L_m(t)$, as shown in Figure 7.2.

As mentioned in CHAPTER 4, the angle θ in the two-dimensional state-space plot of Figure 4.3 does not necessarily represent the phase, $\phi(t)$. A general definition for the phase of the oscillator can be obtained by performing a thought experiment as shown in Figure 7.3. Consider two identical oscillators in which all the noise sources can be turned off using a switch (*i.e.*, $b(t)$ can be set to zero). Both oscillators are started at the time reference $t=t_0$ with identical initial conditions. The noise switch on OSC1 (which will be called the reference oscillator), will be left open at all times. The noise switch on oscillator OSC2 will be on initially. Although they start from the same initial conditions, the waveform of OSC2 will deviate from the ideal waveform of OSC1 due to the perturbations. Now if the noise switch for OSC2 is turned off at time $t=t_1^-$, there is a unique number, $\phi(t_1)$, in (2.1) that will make the *N*th transition of the output of OSC2 occur at exactly the same time as the *N*th transition of the ideal OSC1 for $t \gg t_1$. This value is defined as the instantaneous phase of the oscillator at t_1, *i.e.*, $\phi(t_1)$.

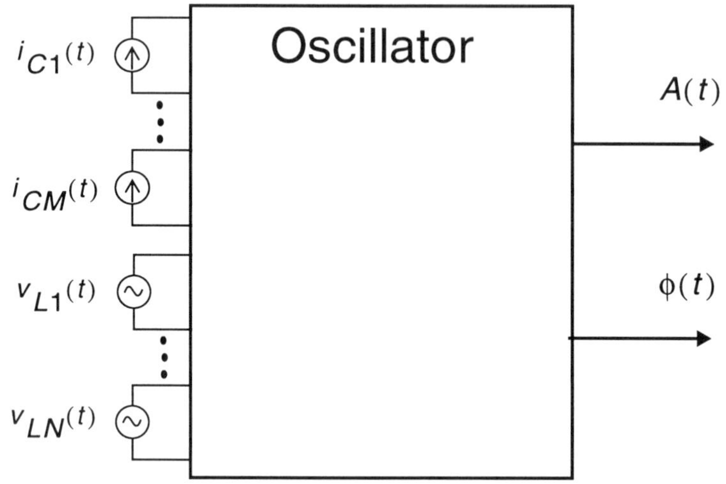

FIGURE 7.2 Oscillator with current and voltage input noise sources.

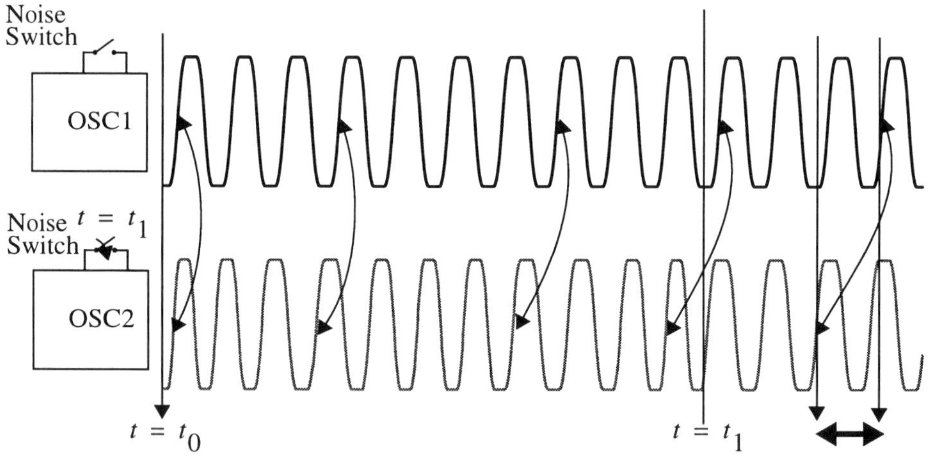

FIGURE 7.3 General definition of phase.

As can be seen from the foregoing discussion, all the state variables undergo the same final phase shift, therefore $\phi(t)$ is the same for *all* of the state variables and is unique for a given perturbation. As can be seen from (2.1), once $\phi(t)$ is defined, $A(t)$ is defined unambiguously. Based on this observation an oscillator can be modeled as a

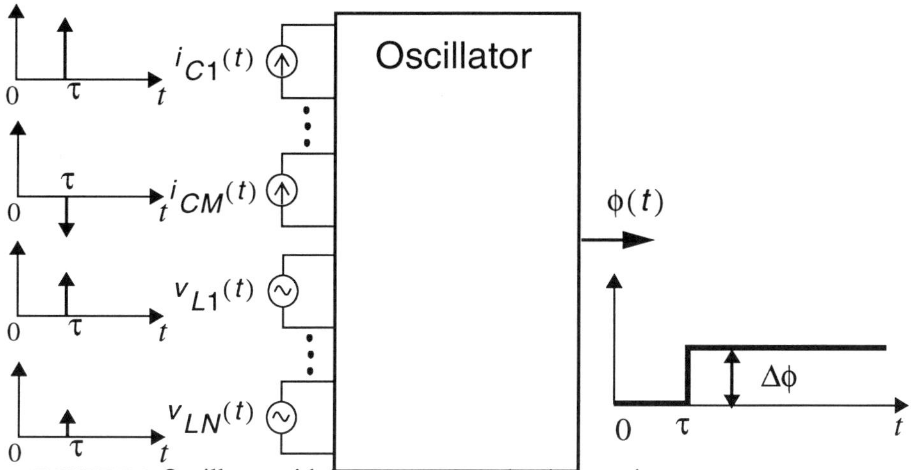

FIGURE 7.4 Oscillator with input current and voltage noise sources.

system with n perturbation sources (elements of $b(t)$) as inputs and $\phi(t)$ as the output, depicted in Figure 7.2.

To characterize the system, a sudden change in its state is considered. In an electrical circuit, this perturbation corresponds to an instantaneous change in capacitor voltages and inductor currents, which can be caused by current impulses in parallel with the capacitors and voltage impulses in series with the inductors as shown in CHAPTER 4. This change can be represented by the vector ΔX in the state-space, which corresponds to $b(t) = \Delta X\delta(t)$, where $\delta(t)$ is the Dirac delta function. This input puts the system in a new state $X + \Delta X$, instantaneously. After application of this perturbation, a stable oscillator undergoes some transient behavior, but finally returns to its limit-cycle. The amplitude variations decay to zero but there will be a net offset in the phase of *all* of the state variables. Thus, the system undergoes a change from the steady-state solution, $X_{\phi_1}(t)$, to another steady-state solution, $X_{\phi_2}(t)$, as shown in Figure 7.1. As mentioned in CHAPTER 4, the phase offset, $\Delta\phi = \phi_1 - \phi_2$, will persist indefinitely since the system has no memory of its previous state. Based on the definition for phase given in this chapter, $\phi(t)$ due to an impulse input will be an ideal step with an amplitude of $\Delta\phi$, as shown in Figure 7.4.

An impulsive perturbation current source in parallel with a capacitor results in an instantaneous change in its voltage. Similarly, an impulsive voltage source in series with an inductor causes a sudden change in its current. The combination of all these impulses at time $t=\tau$ results in an instantaneous phase shift, $\Delta\phi$, as shown in

Figure 7.4. This instantaneous phase shift is a nonlinear function of the areas of the input impulses for sufficiently large areas of input impulses. Calling the areas of the ith current and jth voltage impulses Δq_i and Δp_j, respectively, the resultant output phase shift can be written as

$$\Delta\phi = w(\Delta q_1 \ldots \Delta q_M ; \Delta p_1 \ldots \Delta p_M) = w(\Delta \boldsymbol{X}) \tag{7.7}$$

where w is a function of $M+N$ variables, or equivalently vector $\Delta \boldsymbol{X}$. This function can be expanded in a Taylor series as:

$$\Delta\phi = \sum_{i=1}^{M} \Delta q_i \frac{\partial w}{\partial \Delta q_i} + \sum_{i=1}^{N} \Delta p_i \frac{\partial w}{\partial \Delta p_i} + O^2(\Delta p, \Delta q) = \Delta \boldsymbol{X} \cdot \nabla_{\Delta p, \Delta q} w + O^2(\Delta \boldsymbol{X}) \tag{7.8}$$

where $\nabla_{\Delta p, \Delta q}$ represents the gradient with respect to all Δp's and Δq's, and $O^2(\Delta \boldsymbol{X})$ is of order 2 or higher in Δp's and Δq's. Therefore, for small perturbations, the higher order terms can be ignored, and the system can be treated as a linear system. This concept is depicted in Figure 7.5 which shows the excess phase vs. areas of the impulsive inputs in Δp and Δq as a surface in a three dimensional graph for a hypothetical oscillator. The foregoing linearization is equivalent to approximating this curved surface with a plane for small Δp's and Δq's. This assumption will be valid throughout this treatment since the noise sources of interest are many orders of magnitude smaller than the signal voltages and currents.

Note that this small-perturbation assumption does *not* mean linearization of the nonlinear V-I characteristics of the active elements in the circuit. Linearity of the current-to-excess phase transfer function does not necessarily require the current-to-current (or current-to-voltage) transfer function of the active devices to be linear. Thus, the nonlinearities essential to defining the limit cycle are preserved in this model.

Considering the periodically time-varying nature of the system of Figure 7.4, it can be completely characterized using the time-variant vector phase impulse response vector, $\boldsymbol{h}_\phi(t, \tau)$, whose elements are the phase impulse responses for a unit impulse input on the corresponding input perturbation source. The first cross term in (7.8) is $\sum\sum \Delta p_i \Delta q_j \frac{\partial^2 w}{\partial \Delta p_i \partial \Delta q_j}$, which is second order and hence very small for small Δp's and Δq's. Accordingly, linearity allows us to evaluate the effect of simultaneous changes in multiple state-variables by superposing their individual impacts. The ith element of $\boldsymbol{h}_\phi(t, \tau)$ represents the phase impulse response of the system to the perturbation source $b_i(t)$. The phase impulse response vector, $\boldsymbol{h}_\phi(t, \tau)$, can thus be written as

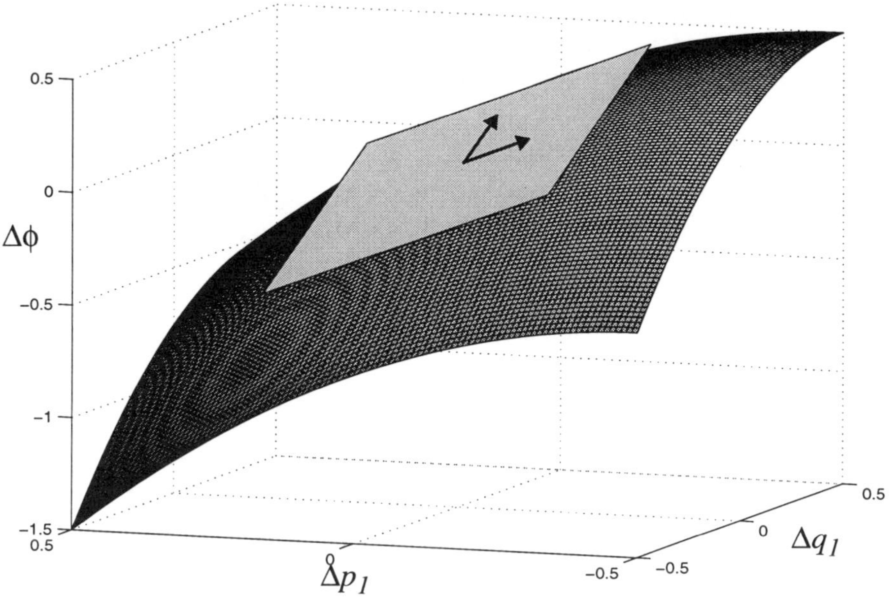

FIGURE 7.5 Phase shift versus the areas of two impulsive perturbations for a hypothetical oscillator.

$$h_\phi(t, \tau) = g(\omega_0\tau)u(t-\tau) \tag{7.9}$$

where $g(x)$ is periodic in x with a period of 2π and can be written as

$$g(\omega_0\tau) = \begin{pmatrix} \Gamma_1(\omega_0\tau)/[C_1(\tau)\cdot V_1] \\ \Gamma_2(\omega_0\tau)/[L_2(\tau)\cdot I_2] \\ \dots \\ \Gamma_n(\omega_0\tau)/[C_n(\tau)\cdot V_n] \end{pmatrix} \tag{7.10}$$

where V_i is the peak voltage across the capacitor C_i and I_j represents the peak current through the inductor L_j. Function $\Gamma_i(\omega_0\tau)$ is the ISF associated with the ith source, as defined in CHAPTER 4.

The time-variant capacitors and inductors in (7.10) make it possible to take into account the effect of nonlinear voltage dependent capacitors and current dependent inductors. If the capacitors and inductors are time-invariant, (7.10) reduces to

$$g(\omega_0\tau) = \begin{pmatrix} \Gamma_1(\omega_0\tau)/q_{1,\,max} \\ \Gamma_2(\omega_0\tau)/p_{1,\,max} \\ \dots \\ \Gamma_n(\omega_0\tau)/q_{k,\,max} \end{pmatrix} \qquad \textbf{(7.11)}$$

where $q_{i,\,max}$ and $p_{j,\,max}$ represent the maximum electric charge in capacitor C_i and maximum magnetic flux in inductor L_j, respectively.

Based on the phase definition in this chapter, the phase shifts introduced at different times add directly[1]. Hence, knowing the impulse response vector, the superposition integral can be applied to calculate the phase at time t. Assuming a zero phase reference at time $-\infty$, phase can be written as

$$\phi(t) = \int_{-\infty}^{\infty} \boldsymbol{b}^T(\tau)\boldsymbol{h}_\phi(t,\tau)d\tau = \int_{-\infty}^{t} \boldsymbol{b}^T(\tau)\boldsymbol{g}[\omega_0\tau + \phi(t)]d\tau \qquad \textbf{(7.12)}$$

where $\boldsymbol{b}^T$ is the transpose of the vector $\boldsymbol{b}$. As argued in CHAPTER 4, for most practical oscillators, the variations in $\phi(t)$ due to noise sources are very small compared to $\omega_0 t$ and therefore (7.12) can be approximated as

$$\phi(t) \approx \int_{-\infty}^{t} \boldsymbol{b}^T(\tau)\boldsymbol{g}(\omega_0\tau)d\tau \qquad \textbf{(7.13)}$$

Note that the phase due to an arbitrary input noise source can be calculated using the superposition integral because of the linearity of the system for small perturbation. If the small perturbation assumption is violated, (7.13) loses its validity. However, as argued in this chapter and shown in CHAPTER 4 for two examples, this assumption holds to an excellent degree for all practical noise sources. The overall process of computing output phase can be modeled as shown in Figure 7.6, where the multiplication of two vectors corresponds to their inner (dot) product.

1. Again this approximation is strictly valid for small perturbations that do not cause a strong deviation from steady-state in the waveform. This assumption is valid to an excellent degree in practical oscillators.

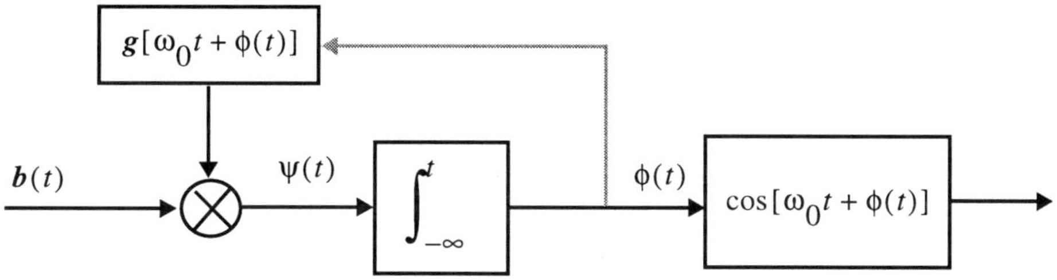

FIGURE 7.6 The equivalent systems for multiple noise sources.

In a fashion similar to CHAPTER 4, $g(\omega_0 t)$ can be written in its Fourier series expansion

$$g(t) = \sum_{n=0}^{\infty} c_n \cos(n\omega_0 t + \theta_n) \tag{7.14}$$

where c_n are constant vectors of Fourier coefficients. Using (7.12) and (7.14)

$$\phi(t) = \sum_{n=0}^{\infty} \int_{-\infty}^{t} b^T(\tau) c_n \cos(n\omega_0 \tau + \theta_n) d\tau \tag{7.15}$$

It can be seen from (7.15) that the frequency components of the input perturbations in the vicinity of integer multiples of the oscillation frequency play the most important role in determining the total $\phi(t)$.

7.2 Phase Noise due to Multiple Stochastic Noise Processes

In this section, the general approach presented in the last section is applied to the case of stochastic noise in actual devices. It starts with a white cyclostationary excitation vector, $b(t)$, and then extends the results to low frequency $1/f$ noise sources.

The output phase noise spectrum obtained in Chapter 3 (and modified in Appendix B) can be generalized to model multiple cyclostationary noise sources with arbitrary correlations, which can be represented as the vector $\boldsymbol{b}(t)$.

Because these cyclostationary processes have the same periodicity as the oscillator waveforms, they are totally correlated with the state variables. A time shift in node voltages and branch currents will therefore result in the same shift in the cyclostationary sources, since they are governed by the same voltage and currents. To evaluate their effect quantitatively, an arbitrary cyclostationary noise vector $\boldsymbol{b}(t)$ can be expressed as

$$\boldsymbol{b}(t) = \tilde{\alpha}(\omega_0 t)\boldsymbol{b}_0(t) \tag{7.16}$$

where $\boldsymbol{b}_0(t)$ is a stationary noise vector and $\tilde{\alpha}(\omega_0 t)$ is a periodic, deterministic diagonal $n \times n$ matrix whose elements modulate the stationary noise vector $\boldsymbol{b}_0(t)$ to give the cyclostationary vector, $\boldsymbol{b}(t)$. Note that $\tilde{\alpha}(\omega_0 t)$ is the natural extension of the NMF defined in CHAPTER 4 and is therefore called *noise modulating matrix.*

The correlation matrix for input source vector $\boldsymbol{b}(t)$ is defined as [138][139]

$$\tilde{\boldsymbol{R}}_b(t_1, t_2) = E[\boldsymbol{b}(t_1)\boldsymbol{b}^T(t_2)] \tag{7.17}$$

Applying (7.16) to (7.17) the correlation matrix can be written as

$$\tilde{\boldsymbol{R}}_b(t_1, t_2) = \tilde{\alpha}(\omega_0 t_1)\tilde{\boldsymbol{R}}_0(t_2 - t_1)\tilde{\alpha}^T(\omega_0 t_2) \tag{7.18}$$

where $\tilde{\boldsymbol{R}}_0(\tau)$ is the correlation matrix of the stationary noise vector, $\boldsymbol{b}_0(t)$.

For multiple noise sources, (7.12) gives $\phi(t)$ as a function of time. Function $\psi(t)$ is defined as

$$\psi(t) = \boldsymbol{b}^T(\tau)\boldsymbol{g}(\omega_0\tau) \tag{7.19}$$

which is a cyclostationary random process because it is the sum of products of stationary random processes and periodic deterministic signals. Equation (7.13) implies that $\psi(t)$ undergoes an ideal integration to give $\phi(t)$:

$$\phi(t) = \int_{-\infty}^{t} \psi(\tau)d\tau \tag{7.20}$$

Finally $\phi(t)$ goes through a phase modulation, as described by Figure 7.6.

The autocorrelation function of $\psi(t)$ is

$$R_{\psi}(t_1, t_2) = E[\psi(t_1)\psi(t_2)] = E\{[\boldsymbol{b}^T(t_1)\boldsymbol{g}(\omega_0 t_1)][\boldsymbol{b}^T(t_2)\boldsymbol{g}(\omega_0 t_2)]\} \tag{7.21}$$

which can be written as

$$\begin{aligned} R_{\psi}(t_1, t_2) &= \boldsymbol{g}^T(\omega_0 t_1)\tilde{\alpha}^T(\omega_0 t_1)\tilde{\boldsymbol{R}}_0(t_2 - t_1)\tilde{\alpha}(\omega_0 t_2)\boldsymbol{g}(\omega_0 t_2) \\ &= \boldsymbol{g}_{eff}^T(\omega_0 t_1)\tilde{\boldsymbol{R}}_0(t_2 - t_1)\boldsymbol{g}_{eff}(\omega_0 t_2) \end{aligned} \tag{7.22}$$

where $\boldsymbol{g}_{eff}$ is defined as

$$\boldsymbol{g}_{eff}(\omega_0 t) = \tilde{\alpha}(\omega_0 t)\boldsymbol{g}(\omega_0 t) \tag{7.23}$$

This result can also be written as

$$R_{\psi}(t_1, t_2) = tr[\tilde{\boldsymbol{G}}(t_1, t_2)\tilde{\boldsymbol{R}}_0(t_2 - t_1)] \tag{7.24}$$

where *tr*[ ] designates the trace (sum of diagonal elements) of a matrix, and the matrix $\tilde{\boldsymbol{G}}(t_1, t_2)$ is defined as

$$\tilde{\boldsymbol{G}}(t_1, t_2) = \boldsymbol{g}_{eff}(\omega_0 t_1)\boldsymbol{g}_{eff}^T(\omega_0 t_2) \tag{7.25}$$

In the case of white noise sources, the correlation matrix can be expressed as

$$\tilde{\boldsymbol{R}}_0(\tau) = \frac{\tilde{\boldsymbol{N}}_0}{2}\delta(\tau) \tag{7.26}$$

where $\tilde{\boldsymbol{N}}_0$ is the noise spectral density matrix determining the noise power and correlations between sources. Therefore,

$$R_{\psi}(t_1, t_2) = \frac{1}{2}tr[\tilde{G}(t_1, t_2)\tilde{N}_0]\delta(t_2 - t_1) \tag{7.27}$$

Defining $\Delta\phi_{\tau}$ as

$$\Delta\phi_{\tau}(t) = \int_{t}^{t+\tau} \psi(t_1)dt_1 \tag{7.28}$$

The variance of $\Delta\phi_{\tau}$ is given by

$$\sigma_{\Delta\phi}^2 = \iint_{t}^{t+\tau} R_{\psi}(t_1, t_2)dt_1 dt_2 = \int_{t}^{t+\tau} \frac{1}{2}tr[\tilde{G}(t_1, t_1)\tilde{N}_0][u(t_1 - t) - u(t_1 - t - \tau)]dt_1 \tag{7.29}$$

Therefore, as shown in Appendix B, for $\tau \gg T$ or $\tau = nT$, the variance is

$$\sigma_{\Delta\phi}^2 = \frac{1}{2}tr[\langle\tilde{G}\rangle\tilde{N}_0] \cdot \tau \tag{7.30}$$

where $\langle\tilde{G}\rangle$ is the time average value of the deterministic matrix $\tilde{G}(t, t)$. In a manner similar to that in Appendix B, the autocorrelation function for the output voltage is given by

$$R_V(\tau) = E[V(t_1)V(t_2)] = \frac{V_0^2}{2}\cos(\omega_0\tau)\exp\left(-\frac{tr[\langle\tilde{G}\rangle\tilde{N}_0]}{4}|\tau|\right) \tag{7.31}$$

Taking the Fourier transform of this autocorrelation function, the following power spectrum for the single sideband phase noise is obtained:

$$\mathcal{L}\{\Delta\omega\} = \frac{8}{tr[\langle\tilde{G}\rangle\tilde{N}_0]} \cdot \frac{1}{1 + \left(\frac{4}{tr[\langle\tilde{G}\rangle\tilde{N}_0]}\right)^2 \Delta\omega^2} \tag{7.32}$$

which is a Lorentzian with a -3dB corner of

$$\omega_{-3dB} = \frac{tr[\langle\tilde{G}\rangle\tilde{N}_0]}{4} \tag{7.33}$$

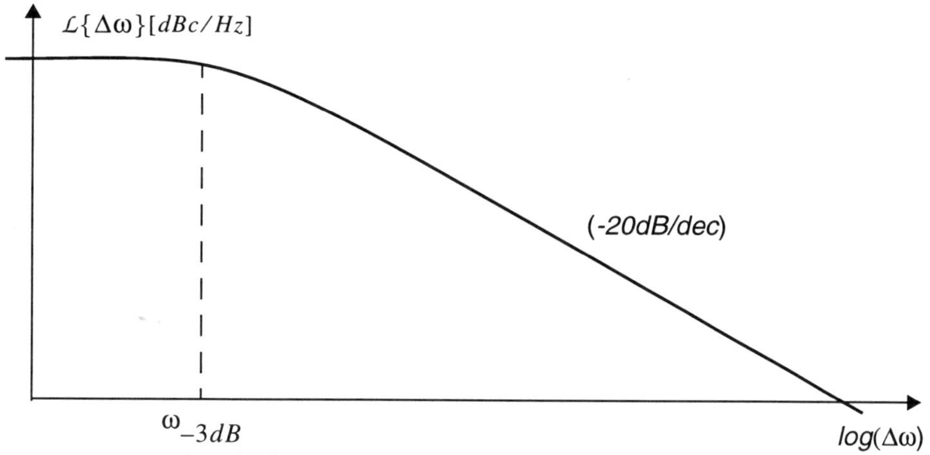

FIGURE 7.7 A typical phase noise plot the Lorentzian spectrum.

as shown hypothetically in Figure 7.7. This Lorentzian spectrum is in accordance with the predicted Lorentzian behavior in [15][16][21][27] and [30]-[32].

A very important point is that although the spectrum of the output voltage of the oscillator follows (7.32), the aforementioned -3dB corner in the phase noise spectrum may or may not be observed, depending on the method used to measure the phase noise. If a spectrum analyzer is directly used to measure the phase noise spectrum through the procedure outline in Section 2.2.1 and Figure 2.7, the -3dB corner given by (7.33) will be seen. If it were possible to use an *ideal* phase detector and a phase locked loop to downconvert the spectrum of $\phi(t)$ and measure it directly, no flattening in the noise spectrum would be seen since the output in that case would be the spectrum of $S_{\phi}(\omega)$. However, since any phase detector will truncate the input phase at some integer multiple of π, the implicit nonlinearity will limit the output power and lead to a -3dB corner frequency *different* from the corner given by (7.33). The same is true if a delay line is used to downconvert $\phi(t)$. This lack of consistency in measurement techniques has been the source of much confusion in the past.

For $\Delta\omega \gg \omega_{-3dB}$, the phase noise spectrum of (7.32) simplifies to

$$\mathcal{L}\{\Delta\omega\} = \frac{tr[\langle\tilde{G}\rangle\tilde{N}_0]}{2\Delta\omega^2} \quad (7.34)$$

For uncorrelated sources, the correlation matrix, $\tilde{N}_0$, will be diagonal; therefore, only the diagonal terms of $\tilde{G}$ appear in the trace operation. Note that the diagonal terms of the correlation matrix are equal to the noise spectral densities of the individual noise sources, and the diagonal terms of $\tilde{G}$ are the rms values of the ISF for each state variable.

In the case of uncorrelated noise sources, the effect of $1/f$ noise can be taken into account by defining the diagonal matrix of corner frequencies of the noise sources, *i.e.,* $\tilde{\Omega}$. In this case, the phase noise in the $1/f^3$ region is given by

$$\mathcal{L}_{1/f^3}\{\Delta\omega\} = \frac{tr[\tilde{D}\tilde{N}_0\tilde{\Omega}]}{2\Delta\omega^3} \tag{7.35}$$

where

$$\tilde{D} = \langle g_{eff}\rangle\langle g_{eff}{}^T\rangle \tag{7.36}$$

Therefore, the $1/f^3$ corner of the phase noise is given by

$$\omega_{1/f^3} = \frac{tr[\tilde{D}\tilde{N}_0\tilde{\Omega}]}{tr[\langle\tilde{G}\rangle\tilde{N}_0]} \tag{7.37}$$

7.3 Summary

The phase impulse response introduced in CHAPTER 4 was generalized in this chapter to accommodate multiple noise sources. Using the generalized impulse response of an oscillator, its output spectrum in the presence of multiple noise sources with arbitrary correlation and cyclostationarity was calculated and a general expression for the phase noise was derived.

CHAPTER 8 Conclusion

A new approach for modeling of jitter and phase noise in electrical oscillators was presented in this work. It is based on modeling an oscillator as two cascaded systems. The first system converts input current and voltage perturbations to phase and is shown to be linear for small perturbations, as well as time-variant. This system is fully characterized by its time-variant impulse response, which can be expressed in terms of charge swing (or flux swing), and a dimensionless impulse sensitivity function (ISF). Superposition allows one to calculate the output phase due to an arbitrary input perturbation using the impulse response. The second system in the cascade is a nonlinear system that converts phase to voltage and whose effect was analyzed in detail. Using these equivalent systems, the mechanism by which device and circuit noise becomes phase noise was demonstrated.

Upconversion of low frequency noise, such as $1/f$ noise, can be understood in this framework. It was shown that the upconversion is governed by the dc value of the ISF, which in turn has a strong dependence on the rise and fall time symmetry of the waveform. It was experimentally demonstrated that the rise and fall time symmetry directly influences the upconversion of the low frequency noise.

The time variant nature of the approach allows it to model properly periodically time-varying effects, such as cyclostationary noise and voltage dependence of capacitors correctly. The cyclostationary noise sources which exist in almost all oscillators arise from the dependence of the noise power spectrum on time varying device currents and voltages. It was shown that the impact of cyclostationary noise sources on the phase noise of the oscillator depends greatly on their relative timing with respect to the ISF.

Application of the model to ring oscillators is simplified by the development of a closed-form approximate expression for the rms and dc values of the ISF. These expressions allow calculation of the lower limits for phase noise and jitter in terms of the number stages, power dissipation, operation frequency, and process parameters. These lower bounds indicate different dependencies on the number of stages for single-ended and differential ring oscillators. Substrate and supply noise sources can result in strongly cross-correlated noise sources on different nodes of the ring oscillator. It was shown that ring oscillators can be designed so that only the noise components in the vicinity of integer multiples of N times[1] the frequency of oscillation have a dominant effect on the phase noise. The effect of substrate and supply noise on jitter and phase noise can therefore be reduced if such perturbations can be made common-mode.

The phase noise of cross-coupled CMOS LC oscillators was analyzed and the contribution of various noise sources to the phase noise was calculated. A simple expression for the tank amplitude of such oscillators was obtained and used to estimate phase noise. Also a straightforward method for exploiting the cyclostationary properties of the noise sources was demonstrated.

The model was generalized to multiple noise sources with arbitrary cross-correlation and cyclostationarity, by defining a generalized impulse response vector. Using this impulse response vector, a new general expression for the phase noise in the presence of multiple noise sources was presented.

1. N is the number of stages.

APPENDIX A

Relationship between Jitter and Phase Noise

Analog and digital designers prefer using phase noise and timing jitter, respectively. The relationship between these two parameters can be obtained by noting that timing jitter is the standard deviation of the timing uncertainty:

$$\sigma_\tau^2 = \frac{1}{\omega_0^2}E\{[\phi(t+\tau)-\phi(t)]^2\} = \frac{E[\phi^2(t)]}{\omega_0^2} + \frac{[\phi^2(t+\tau)]}{\omega_0^2} - \frac{2E[\phi(t)\phi(t+\tau)]}{\omega_0^2} \quad \text{(A.1)}$$

where $E[.]$ represents the expected value. Since the autocorrelation function of $\phi(t)$, $R_\phi(\tau)$, is defined as

$$R_\phi(\tau) = E[\phi(t)\phi(t+\tau)] \quad \text{(A.2)}$$

the timing jitter in (A.1) can be written as

$$\sigma_\tau^2 = \frac{2}{\omega_0^2}[R_\phi(0) - R_\phi(\tau)] \quad \text{(A.3)}$$

The relation between the autocorrelation and the power spectrum is given by the Khinchin theorem, *i.e.*, [138]

inverse Fourier

$$R_\phi(\tau) = \frac{1}{2\pi}\int_{-\infty}^{\infty} S_\phi(\omega)e^{j\omega\tau}d\omega \qquad \textbf{(A.4)}$$

where $S_\phi(\omega)$ represents the power spectrum of $\phi(t)$. Therefore, (A.3) results in the following relationship between clock jitter and phase noise:

$$\sigma_\tau^2 = \frac{4}{\pi\omega_0^2}\int_0^{\infty} S_\phi(\omega)sin^2\left(\frac{\omega\tau}{2}\right)d\omega \qquad \textbf{(A.5)}$$

It may be useful to know that $S_\phi(\omega)$ can be approximated by $\mathcal{L}\{\Delta\omega\}$ defined in CHAPTER 2 for large offsets [66].

As can be seen from the foregoing, the rms timing jitter has less information than the phase noise spectrum and can be calculated from phase noise using equation (A.5). However, unless extra information about the shape of the phase noise spectrum is known, the inverse is not possible in general.

In the special case where the phase noise is dominated by white noise, $\mathcal{L}\{\Delta\omega\}$ and κ are given by (4.16) and (4.23). Therefore, κ can be expressed in terms of phase noise in the $1/f^2$ region as:

$\mathcal{L}\{\Delta\omega\} = -20\log_{10}\left(K\frac{\omega}{\Delta\omega}\right)$

$\mathcal{L}\{\Delta\omega\} = -10\log_{10}\left(K^2\frac{\omega^2}{\Delta\omega^2}\right)$

$$\kappa = \frac{\Delta\omega}{\omega_0}\cdot 10^{-\mathcal{L}\{\Delta\omega\}/20} \qquad \textbf{(A.6)}$$

where $\mathcal{L}\{\Delta\omega\}$ is the phase noise measured in the $1/f^2$ region at an offset frequency of $\Delta\omega$ and ω_0 is the oscillation frequency. Therefore, based on (2.6), the cycle-to-cycle jitter will be given by

$$\sigma_{CTC} = \sqrt{\frac{2\pi}{\omega_0}}\cdot\frac{\Delta\omega}{\omega_0}\cdot 10^{-\mathcal{L}\{\Delta\omega\}/20} \qquad \textbf{(A.7)}$$

Note that for (A.6) and (A.7) to be valid, the phase noise at $\Delta\omega$ should be in the $1/f^2$ region.

APPENDIX B

Power Spectral Density of the Output

The output spectrum of an oscillator due to noise was originally analyzed in [14]. Shortly thereafter, the problem was analyzed in more detail to show that the output spectrum in the presence of amplitude and phase fluctuations is Lorentzian [15][16]. These approaches were extended to relate the time and frequency domain representations of frequency instabilities [17]-[23]. The approaches of [14]-[20] were generalized and applied to optical systems using a stochastic noise approach [21], showing that phase undergoes a random walk process.

It is shown in this appendix that for the time variant phase noise model of CHAPTER 4, the output power spectrum of an oscillator with a single white input noise source is Lorentzian in agreement with previous observations. According to (4.22), for a single cyclostationary noise source *i(t)*, the phase jitter after τ seconds is given by:

$$\sigma_{\Delta\phi}^2 = E[\Delta\phi_\tau^2(t)] \cong \frac{\Gamma_{rms}^2 \cdot \overline{i_n^2}/\Delta f}{2q_{max}^2}|\tau| \qquad \begin{array}{c} \tau \gg T \\ or \\ \tau = nT \end{array} \tag{B.1}$$

Excess phase, $\phi(t)$, eventually undergoes a phase modulation (PM) as illustrated in Figure 4.11. This process is characterized by (2.1). Assuming negligible amplitude fluctuations[1] and looking at the fundamental component of (2.1), the autocorrelation of $V(t)$ is given by

$$R_V(t_1, t_2) = E[V(t_1)V(t_2)]$$
$$= V_0^2 \cdot E\{\cos[\omega_0 t_1 + \phi(t_1)]\cos[\omega_0 t_2 + \phi(t_2)]\} \quad \textbf{(B.2)}$$

which can be written as

$$R_V(t_1, t_2) = \frac{V_0^2}{2}E\{\cos[\omega_0(t_2 - t_1) + \phi(t_2) - \phi(t_1)]\} +$$
$$\frac{V_0^2}{2}E\{\cos[\omega_0(t_1 + t_2) + \phi(t_1) + \phi(t_2)]\} \quad \textbf{(B.3)}$$

The central limit theorem indicates that $\Delta\phi_\tau$ has a Gaussian probability density function [39], *i.e.*,

some type of random walk result

$$p(\Delta\phi_\tau) = \frac{1}{\sigma_{\Delta\phi}\sqrt{2\pi}}\exp\left(-\frac{\Delta\phi_\tau^2}{2\sigma_{\Delta\phi}^2}\right) \quad \textbf{(B.4)}$$

Hence,

$$E[\cos\Delta\phi_\tau] = \int_{-\infty}^{\infty} p(\Delta\phi_\tau)\cos(\Delta\phi_\tau)d(\Delta\phi_\tau) = \exp\left(-\frac{\sigma_{\Delta\phi}^2}{2}\right) = \exp\left(-\frac{\Gamma_{rms}^2 \cdot \overline{i_n^2}/\Delta f}{4q_{max}^2}|\tau|\right)$$
$$E[\sin\Delta\phi_\tau] = \int_{-\infty}^{\infty} p(\Delta\phi_\tau)\sin(\Delta\phi_\tau)d(\Delta\phi_\tau) = 0 \qquad = \exp(-D_\phi|\tau|) \quad \textbf{(B.5)}$$

Now it will be shown that the second term in (B.3) converges to zero as $t_1 \to -\infty$. The expression of interest, as $t_1 \to -\infty$, is

$$R_{V2}(t_1, t_2) = \frac{V_0^2}{2}E\{\cos[\omega_0(t_1 + t_2) + \phi(t_1) + \phi(t_2)]\} \quad \textbf{(B.6)}$$

which can be written as

1. The case in which both amplitude and phase fluctuation terms exist can also be analyzed in a similar fashion.

$$R_{V2}(t_1, t_2) = \cos\omega_0(t_1 + t_2)E\{\cos[\phi(t_1) + \phi(t_2)]\} - \sin\omega_0(t_1 + t_2)E\{\sin[\phi(t_1) + \phi(t_2)]\} \tag{B.7}$$

Knowing that

$$\phi(t_1) = \int_{t_0}^{t_1} \psi(t_3)dt_3$$
$$\phi(t_2) = \int_{t_0}^{t_2} \psi(t_4)dt_4 \tag{B.8}$$

hence,

$$\phi(t_1) + \phi(t_2) = 2\int_{t_0}^{t_1} \psi(t_3)dt_3 + \int_{t_1}^{t_2} \psi(t_4)dt_4 = 2\phi(t_1) + \Delta\phi_\tau \tag{B.9}$$

The first term of (B.7) can be expanded as

$$\begin{aligned} E\{\cos[\phi(t_1) + \phi(t_2)]\} &= E\{\cos[2\phi(t_1) + \phi_\tau]\} \\ &= E\{\cos[2\phi(t_1)]\cos\Delta\phi_\tau\} - E\{\sin[2\phi(t_1)]\sin\Delta\phi_\tau\} \end{aligned} \tag{B.10}$$

Since ϕ is a Wiener processes, $\phi(t_1)$ and $\Delta\phi_\tau$ are disjoint, and therefore independent. Therefore,

$$E\{\cos[\phi(t_1) + \phi(t_2)]\} = E\{\cos[2\phi(t_1)]\}E\{\cos\Delta\phi_\tau\} - E\{\sin[2\phi(t_1)]\}E\{\sin\Delta\phi_\tau\} \tag{B.11}$$

and from (B.5) following is obtained

$\exp(-D_\phi|\tau|)$

$$E\{\cos[\phi(t_1) + \phi(t_2)]\} = E\{\cos[2\phi(t_1)]\}\exp\left(-\frac{\Gamma_{rms}^2 \cdot \overline{i_n^2}/\Delta f}{4q_{max}^2}|\tau|\right) \tag{B.12}$$

Defining $\tau' = t_2 - t_0$, the expected value of $E\{\cos[2\phi(t_1)]\}$ can be written as

$$E\{\cos[2\phi(t_1)]\} = \int_{-\infty}^{\infty} p(\Delta\phi_{\tau'})\cos(2\Delta\phi_{\tau'})d\Delta\phi_{\tau'} = \frac{1}{2}\exp(-2\sigma^2_{\Delta\phi'})$$

$$= +\frac{1}{2}\exp\left(-\frac{\Gamma^2_{rms}\cdot\overline{i^2_n}/\Delta f}{q^2_{max}}|\tau'|\right) \quad = \frac{1}{2}\exp(-4D_\phi|\tau|) \tag{B.13}$$

and in a similar manner,

$$E\{\sin[2\phi(t_1)]\} = 0 \tag{B.14}$$

Thus,

$$\exp(-D_\phi|\tau|) \qquad \exp(-4D_\phi|\tau'|)$$

$$R_{V2}(t_1, t_2) =$$

$$\lim_{t_0\to-\infty}\frac{1}{2}\cos\omega_0(t_1+t_2)\exp\left(-\frac{\Gamma^2_{rms}\cdot\overline{i^2_n}/\Delta f}{4q^2_{max}}|t_2-t_1|\right)\exp\left(-\frac{\Gamma^2_{rms}\cdot\overline{i^2_n}/\Delta f}{q^2_{max}}|t_2-t_0|\right) = 0 \tag{B.15}$$

which was to be shown.

Based on (B.3) and (B.15), the autocorrelation function of the output is given by

$$R_V(\tau) = \frac{V_0^2}{2}\cos(\omega_0\tau)E[\cos\Delta\phi_\tau] - \frac{V_0^2}{2}\sin(\omega_0\tau)E[\sin\Delta\phi_\tau] \tag{B.16}$$

where $\tau = t_2 - t_1$. Using (B.5), the autocorrelation function for the output voltage (or current) is given by

$$R_V(\tau) = \frac{V_0^2}{2}\cos(\omega_0\tau)\exp\left(-\frac{\Gamma^2_{rms}\cdot\overline{i^2_n}/\Delta f}{4q^2_{max}}|\tau|\right) \tag{B.17}$$

The double-sideband power spectrum can be calculated by taking the Fourier transform of (B.17). The single-sideband phase noise power below the carrier at an offset frequency $\Delta\omega$ is directly calculated from the double sideband spectrum as

$$R_V(\tau) = \frac{V_0^2}{2}\exp(-D_\phi|\tau|)\cos(\omega_0\tau)$$

$$S_V(\omega) = \frac{V_0^2}{2}\;\frac{2D_\phi}{(\omega-\omega_0)^2 + D_\phi^2}$$

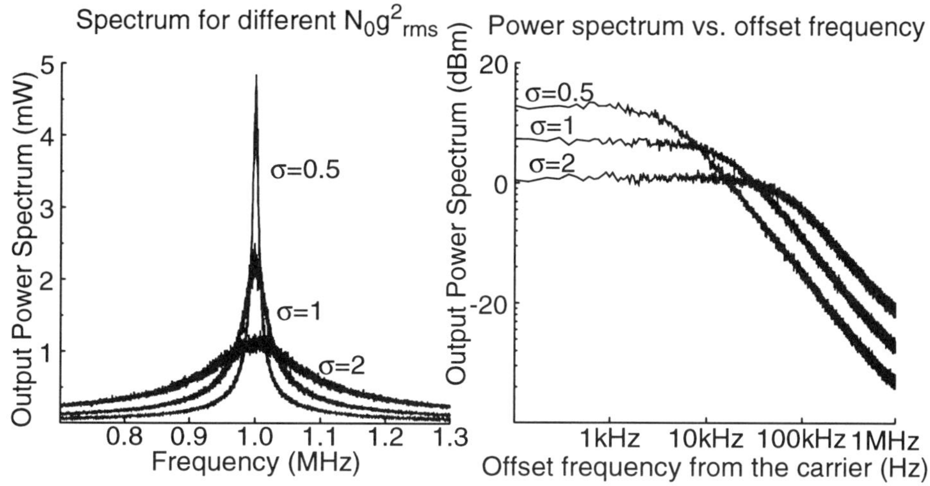

FIGURE B.1 Simulated output power spectrum vs. a) frequency b) offset from the carrier.

$$\mathcal{L}\{\Delta\omega\} = \frac{8q_{max}^2}{\Gamma_{rms}^2 \cdot \overline{i_n^2}/\Delta f} \cdot \frac{1}{1 + \left(\frac{4q_{max}^2}{\Gamma_{rms}^2 \cdot \overline{i_n^2}/\Delta f}\right)^2 \Delta\omega^2} \tag{B.18}$$

which is a Lorentzian with a -3dB corner of

$$\omega_{-3dB} = \frac{\Gamma_{rms}^2 \cdot \overline{i_n^2}/\Delta f}{4q_{max}^2} \tag{B.19}$$

A similar spectrum is obtained by a time domain Monte Carlo simulation and is shown in Figure B.1a and b for three different values of σ and the same V_0. As can be seen, this simulation confirms the result of (B.18). Note that the value of σ has been artificially made extremely large so that its effect can be seen without having to perform very long simulation. In practical LC oscillators, this corner is in the range of a few Hz. As can be seen from (B.19) and Figure B.1, the better the phase noise at a given offset, the lower this corner will be.

Note that the integral of the output power spectrum given by (B.18) does not diverge to infinity, even though the power spectral density of $\phi(t)$ grows without bound as

$\Delta\omega$ approaches zero. The apparently non-physical power spectrum of $\phi(t)$ creates no problem since $\phi(t)$ is not an actual voltage or current which can be measured directly; it can only be indirectly observed in the shift of zero-crossings.

It turns out that the corner frequency seen in Figure B.1b is small for most practical applications. Therefore, for $\Delta\omega \gg \omega_{-3dB}$, (B.18) simplifies to

$$\mathcal{L}\{\Delta\omega\} = \frac{\Gamma_{rms}^2 \cdot \overline{i_n^2}/\Delta f}{2q_{max}^2 \Delta\omega^2} \tag{B.20}$$

which is the same as (4.4).

APPENDIX C

The ISF of an Ideal LC Oscillator

Consider the ideal LC oscillator of Figure C.1. Assuming that the tank has a maximum voltage amplitude of V_0, the voltage across the capacitor and the current through the inductor can be written as

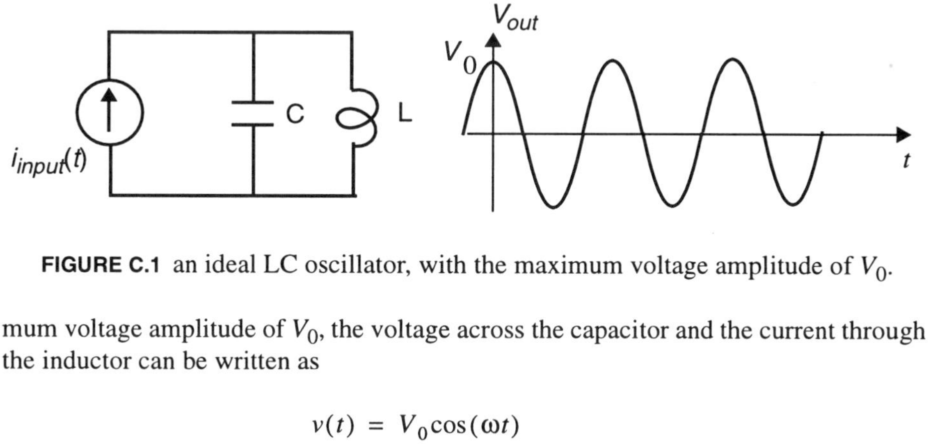

FIGURE C.1 an ideal LC oscillator, with the maximum voltage amplitude of V_0.

$$v(t) = V_0 \cos(\omega t)$$

$$i(t) = V_0 \sqrt{\frac{C}{L}} \sin(\omega t) \tag{C.1}$$

where L and C are the values of the inductor and capacitor, respectively, and $\omega = 1/\sqrt{LC}$ is the angular frequency of oscillation[1].

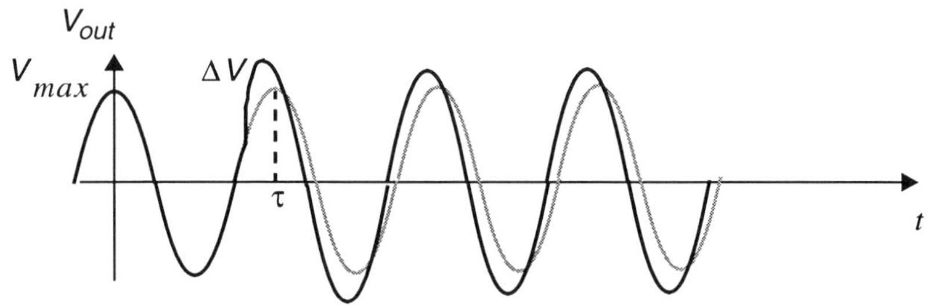

FIGURE C.2 Induced phase and amplitude changes due to a change in the voltage.

If a current impulse with an area of Δq is injected into the tank at $t=t_0$, it will induce a voltage change of $\Delta q/C$ in the capacitor voltage, as shown in Figure C.2. Therefore, the capacitor voltage at t_0^+ is $V_0\cos(\omega t_0) + \Delta q/C$ and the inductor current does not change and is $V_0\sqrt{C/L}\sin(\omega t_0)$. The capacitor voltage and the inductor current after t_0 will be sinusoids with a phase shift, $\Delta\phi$, and an amplitude change, ΔV, with respect to the initial sinusoid, *i.e.*,

$$v(t) = (V_0 + \Delta V)\cos(\omega t + \Delta\phi)$$
$$i(t) = (V_0 + \Delta V)\sqrt{\frac{C}{L}}\sin(\omega t + \Delta\phi) \tag{C.2}$$

The voltage and current given by (C.2) should be equal to the initial condition at t_0^+, *i.e.*,

$$(V_0 + \Delta V)\cos(\omega t_0 + \Delta\phi) = V_0\cos(\omega t_0) + \frac{\Delta q}{C}$$
$$(V_0 + \Delta V)\sin(\omega t_0 + \Delta\phi) = V_0\sin(\omega t_0) \tag{C.3}$$

Expanding the cosine and sine functions (C.3) can be written as

1. Note that a zero initial phase (epoch) is chosen for the voltage and current. The choice of initial phase will be governed by the choice of the time origin and is therefore arbitrary. Since only the induced phase change is evaluated, the initial value of the phase does not matter.

$$(V_0 + \Delta V)[\cos(\omega t_0)\cos(\Delta\phi) - \sin(\omega t_0)\sin(\Delta\phi)] = V_0\cos(\omega t_0) + \frac{\Delta q}{C}$$

$$(V_0 + \Delta V)[\sin(\omega t_0)\cos(\Delta\phi) + \cos(\omega t_0)\sin(\Delta\phi)] = V_0\sin(\omega t_0) \quad \textbf{(C.4)}$$

Since $\Delta\phi$ and ΔV are small, $\cos(\Delta\phi) \approx 1$, $\sin(\Delta\phi) \approx \Delta\phi$ and $V_0 + \Delta V \approx V_0$, are valid approximations. Using these approximations, (C.4) can be written as

$$\Delta V\cos(\omega t_0) - V_0\Delta\phi\sin(\omega t_0) = \frac{\Delta q}{C}$$

$$\Delta V\sin(\omega t_0) + V_0\Delta\phi\cos(\omega t_0) = 0 \quad \textbf{(C.5)}$$

Multiplying the first and second equations in (C.5) by $\sin(\omega t_0)$ and $\cos(\omega t_0)$, respectively, and subtracting the first from the second, the following is obtained

$$\Delta\phi = -\frac{\Delta q}{CV_0}\sin(\omega t_0) \quad \textbf{(C.6)}$$

and, therefore, the phase impulse response is

$$h_\phi(t, \tau) = \frac{-\sin(\omega\tau)}{q_{max}}u(t - \tau) \quad \textbf{(C.7)}$$

Comparing (C.7) with (4.3), the phase ISF for an ideal oscillator is

$$\Gamma(x) = -\sin(x) \quad \textbf{(C.8)}$$

APPENDIX D

Calculation of the ISF

The ISF can be calculated using two different methods. The first method is based on direct injection of an impulse at different times and measurement of induced phase shift. The second method is based on estimation of the impulse response from the steady-state solution. The first method is more accurate, but more computationally intensive.

D.1 Direct Impulse Response Calculation

This method is based on the observation that the most accurate and straightforward way of calculating the ISF is to replace the noise sources in the circuit with impulsive sources of small area and measure the resultant phase shift. Repeating this process by injecting the impulse at various times during a period and measuring the resultant phase shift allows the ISF to be calculated. This method is the most accurate way to obtain the ISF because it makes no limiting assumptions. It is noteworthy that, although the unit impulse response is used, applying an actual impulse with area of 1 coulomb would drive the oscillator deep into nonlinear behavior. Therefore, the unit impulse response is obtained by applying impulses with areas much smaller than the steady-state charge swing into the energy storage element of interest and scaling accordingly.

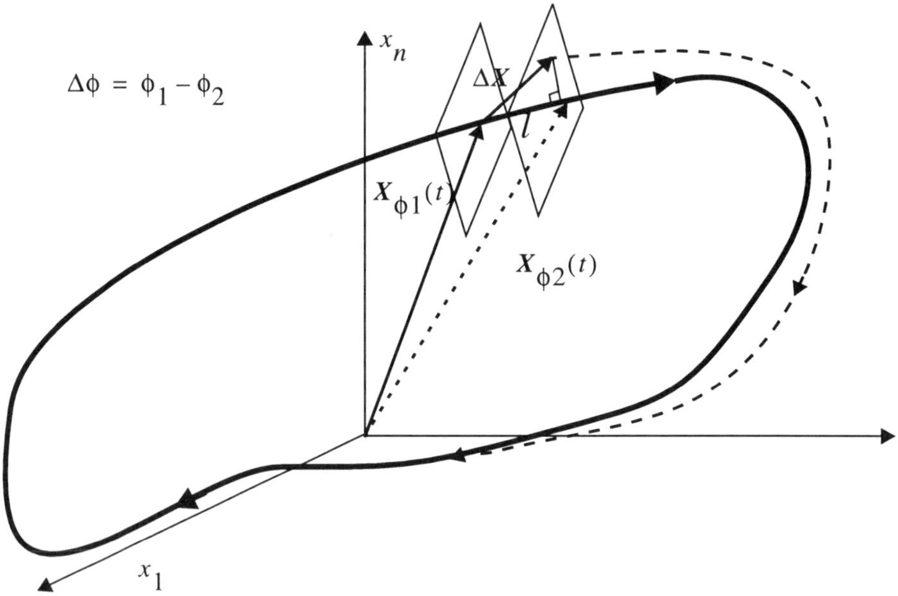

FIGURE D.1 Limit cycle of an arbitrary oscillator in the state-space, showing the effect of an arbitrary perturbation on the phase of the oscillator.

If there are P different noise sources in the circuit, and it is desired to find Q points in one period of the ISF for each source, the transient analysis of the oscillator needs to be repeated PQ times. Each time the oscillator should be simulated to reach the steady-state after injection of the perturbation. The unperturbed oscillator also needs to be simulated once more to obtain the unperturbed response for comparison. Therefore, this method requires one to simulate the oscillator $PQ+1$ times.

D.2 Calculation of the Impulse Response from the Steady-State Solution

This method is based on a different approach which allows quick estimation of the ISF from the steady-state solution. The process starts by normalizing the state variables to their maximum values to obtain dimensionless state-variables. This new state-vector is designated as $\boldsymbol{X}$, and an instantaneous normalized change in the state of the system as $\Delta\boldsymbol{X}$, as shown in Figure D.1. After application of $\Delta\boldsymbol{X}$, the state of the system eventually returns to its steady-state limit cycle, but generally at a different phase, as discussed earlier.

The voltage change can be decomposed into orthogonal and tangential components [31]. This phase shift can be estimated by noting that if the magnitude of this perturbation vector, $\Delta \boldsymbol{X}$, is small, the perturbation in the direction of the motion has the largest effect on the phase shift, while the perturbation orthogonal to the direction of the motion has a much smaller effect on the final phase. The perturbation of the amplitude eventually decays to zero, as implied by the existence of the limit cycle.

Based on the foregoing argument, the final phase shift due to a perturbation vector with an arbitrary direction, $\Delta \boldsymbol{X}$, can be calculated by first calculating the resultant displacement along the trajectory of the state. This is given by the inner product of the perturbation vector and a unit vector in the direction of the motion, *i.e.*,

$$l = \Delta \boldsymbol{X} \cdot \frac{\dot{\boldsymbol{X}}}{|\dot{\boldsymbol{X}}|} \quad \text{(D.1)}$$

where l is the equivalent displacement along the trajectory and $\dot{\boldsymbol{X}}$ is the derivative of the state vector with respect to time, t. Note the scalar nature of l, which arises from the projection operation. Also note that the oscillator does not necessarily traverse the limit cycle with a constant "speed." The resultant phase shift is equal to the normalized time it would have taken the system to reach a state l units ahead on the limit cycle. Hence, the phase shift is given by the displacement divided by the "speed"

$$\Delta\phi = \frac{l}{|\dot{\boldsymbol{X}}|} = \Delta \boldsymbol{X} \cdot \frac{\dot{\boldsymbol{X}}}{|\dot{\boldsymbol{X}}|^2} \quad \text{(D.2)}$$

Therefore $\boldsymbol{g}$ of Chapter 5 can be approximated by

$$\boldsymbol{g} = \frac{\dot{\boldsymbol{X}}}{|\dot{\boldsymbol{X}}|^2} \quad \text{(D.3)}$$

Note that (D.3) neglects AM-to-PM conversion in the oscillator. Although it is possible to take the effect of this AM-to-PM conversion into account [31], neglecting AM-to-PM conversion is not a dominant source of error in prediction of phase noise in integrated electrical oscillators. Nevertheless, if there is interest in modeling of AM-to-PM conversion, as well as other mechanisms, the impulse response may always be calculated directly with the first method.

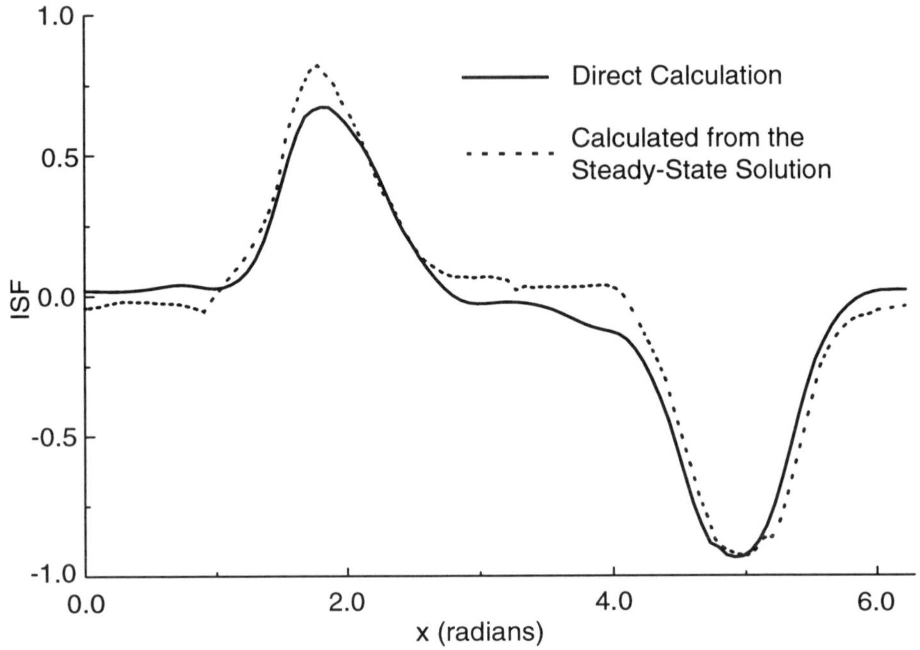

FIGURE D.2 The ISF of the ring oscillator of Figure 4.6 obtained using from different methods.

In the case of an electrical oscillator, the steady-state solution of the circuit can be obtained using circuit simulators, such as SPICE. Note that, to calculate the ISF using the second method, only the steady-state solution in the absence of any perturbation for one cycle is needed. Figure D.2 compares the ISF obtained from the direct calculation method with the ISF calculated from the steady-state solution for the 5 stage ring oscillator of Figure 4.6. As mentioned before the first method results in more accurate predictions since the second order effects such as AM-to-PM conversion is not ignored.

APPENDIX E Phase Noise and Jitter in Phase-Locked Loops

Phase-locked loop (PLL) are an essential block in many applications. PLLs are used in data communication circuits for clock recovery generation, in microprocessors to generate a low skew/jitter clock across the chip, and in RF applications as frequency synthesizers to produce a digitally controllable stable high frequency source from a low frequency reference, such as a crystal oscillator. In this appendix, a brief review of PLL concepts is given and phase noise and jitter behavior of the circuit are discussed.

E.1 A Brief Review of PLLs

Phase-locked loops have been studied extensively in the past. For comprehensive reviews of PLLs, the reader can consult with [64]-[69]. This section starts with an analysis of the first-order loop and then extends this approach to higher order loops and the case of frequency synthesizers.

E.1.1 First Order Loop

Figure E.1 shows the block diagram of a first order phase-locked loop. It consists of a phase detector and voltage controlled oscillator (VCO). In a first-order loop the phase detector is usually implemented using an analog multiplier or an XOR gate which

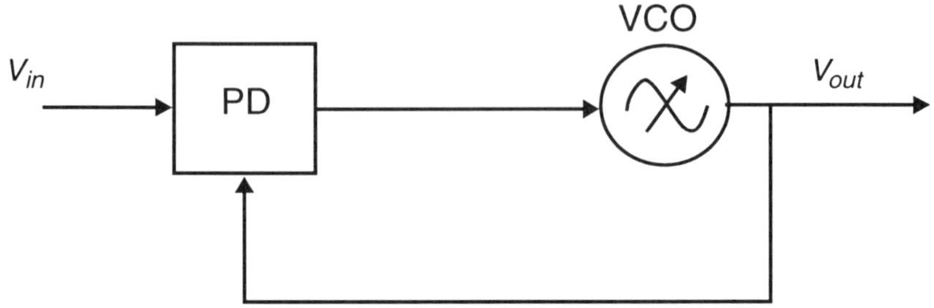

FIGURE E.1 A first order phase locked loop.

compares the phases of the input signal and the output of the VCO. The control voltage of the VCO is provided by the output of the phase detector. Under the locked condition, the negative feedback adjusts the dc value of the VCO control voltage in such a way that it oscillates at the same frequency as the input. When in lock, the input and output will have a constant, known phase relation. Depending on the type of the phase detector used, the constant phase error can be 0 or $\pi/2$.

PLLs are best analyzed in the phase domain. It is instructive to calculate the phase transfer characteristic from the input to the output. In an ideal phase detector, the dc value of the phase detector output is proportional to the phase difference of the input and the output signals. Therefore, it can be modeled as the cascade of a summing node and a constant gain block with a gain of, K_P. The VCO output frequency is proportional to the control voltage. The output frequency of the VCO is proportional to its control voltage. Since phase is the integral of the frequency, the VCO acts as an ideal integrator for the input voltage when the output variable is phase. Therefore, its frequency response can be simply expressed as

$$H_{VCO}(s) = \frac{K_V}{s} \quad \text{(E.1)}$$

In the phase domain, the PLL can be modeled using the equivalent system of Figure E.2. The input-output phase transfer function of the equivalent system of Figure E.2 is

$$\frac{\Phi_{out}(s)}{\Phi_{in}(s)} = \frac{K_P K_V}{K_P K_V + s} \quad \text{(E.2)}$$

with a loop bandwidth of

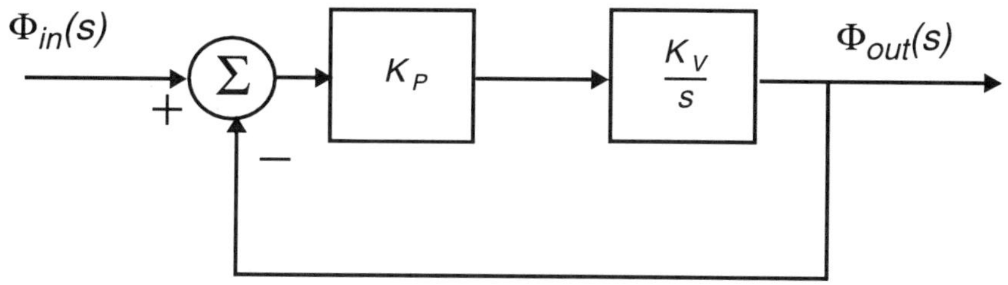

FIGURE E.2 The phase domain equivalent of the first order phase locked loop.

$$\omega_{loop} = K_P K_V \quad \text{(E.3)}$$

The steady-state phase difference between input and output is defined as static error and is easily calculated to be [69]

$$\phi_{error} = \frac{\omega_{in}}{\omega_{loop}} \quad \text{(E.4)}$$

where ω_{in} is the angular frequency of the input. As can be seen, the only way to lower the steady-state phase error is to increase the loop bandwidth.

First order loops are seldom used because of this strong coupling between the bandwidth and the steady-state phase error.

E.1.2 Higher Order Loops

Adding extra degrees of freedom to the loop allows the designer to decouple different loop parameters. The extra degrees of freedom can be added by placing a filter in the forward path between the phase detector and the VCO as shown in Figure E.3. The phase domain equivalent system for the PLL in Figure E.3 is shown in Figure E.4. The transfer function for the system of Figure E.4 is

$$\frac{\Phi_{out}(s)}{\Phi_{in}(s)} = \frac{K \cdot H(s)}{K \cdot H(s) + s} \quad \text{(E.5)}$$

where $K = K_P K_V$. If the PLL is implemented using multipliers or XOR phase detector and an RC low pass filter, the filter transfer function can be written as

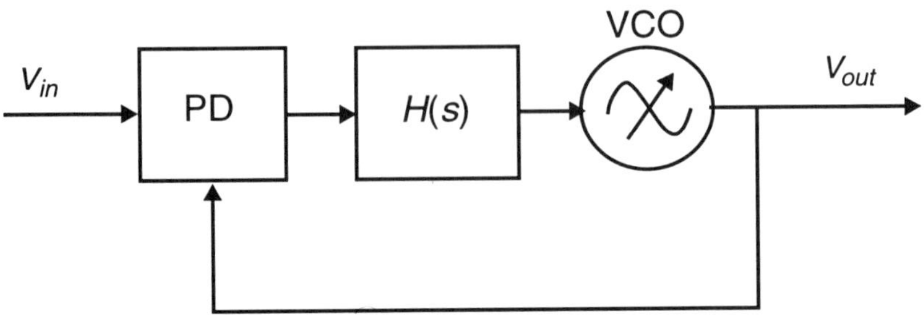

FIGURE E.3 A higher order phase locked loop.

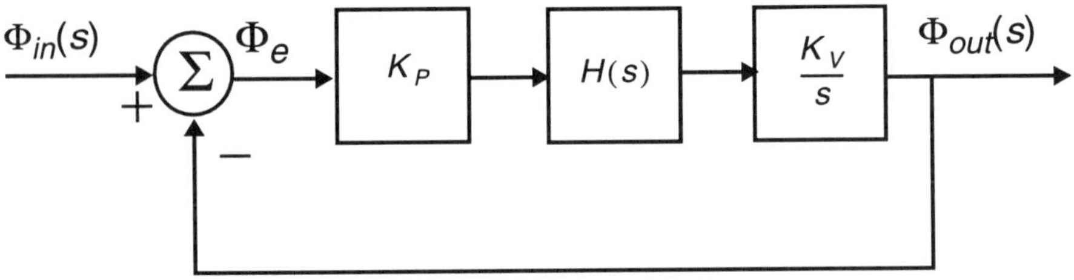

FIGURE E.4 The phase domain equivalent of the higher order phase locked loop.

$$H(s) = \frac{1}{1 + RCs} \tag{E.6}$$

The overall loop transfer function is given by

$$\frac{\Phi_{out}(s)}{\Phi_{in}(s)} = \frac{K\omega_{RC}}{s^2 + \omega_{RC}s + K\omega_{RC}} \tag{E.7}$$

where $\omega_{RC} = 1/RC$. The phase transfer (E.7) can be written in the following more familiar form

$$\frac{\Phi_{out}(s)}{\Phi_{in}(s)} = \frac{\omega_n^2}{s^2 + 2\zeta\omega_n s + \omega_n^2} \tag{E.8}$$

with $\omega_n = \sqrt{K\omega_{RC}}$ and $\zeta = \frac{1}{2}\sqrt{\omega_{RC}/K}$.

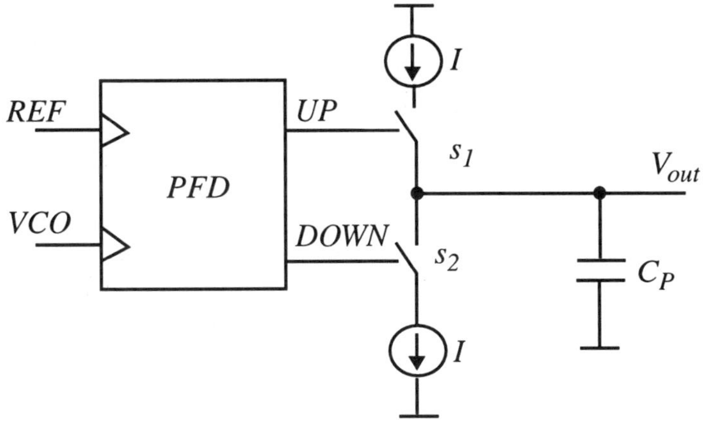

FIGURE E.5 A typical charge pump phase-frequency detector.

The phase error has the following transfer function,

$$\frac{\Phi_e(s)}{\Phi_{in}(s)} = \frac{s^2 + 2\zeta\omega_n s}{s^2 + 2\zeta\omega_n s + \omega_n^2} \tag{E.9}$$

where $\Phi_{in}(s)$ represents the phase error at the input as shown in Figure E.4. If the input is at an angular frequency of ω_i the phase grows linearly and hence $\Phi_{in}(s) = \omega_i/s^2$. The steady-state phase error is

$$\phi_{error} = \lim_{s \to 0} s\Phi_e(s) = \lim_{s \to 0} s\frac{\omega_i}{s^2} \cdot \frac{s^2 + 2\zeta\omega_n s}{s^2 + 2\zeta\omega_n s + \omega_n^2} = \frac{\omega_i}{K} \tag{E.10}$$

which is the same as the static error of the first order loop. Subsequently, the RC loop filter does not eliminate the static phase error.

The static phase error can be eliminated by introduction of a pole at the origin. This corresponds to an ideal integration in the forward path and can be implemented using the charge pump phase-frequency detector (PFD) and a charge pump shown in Figure E.5.

In the charge pump architecture, the PFD has two edge sensitive inputs and two outputs called *UP* and *DOWN*. If the VCO runs at a lower frequency than the input, the *UP* signal will be non-zero intermittently, while the *DOWN* pulse will be zero contin-

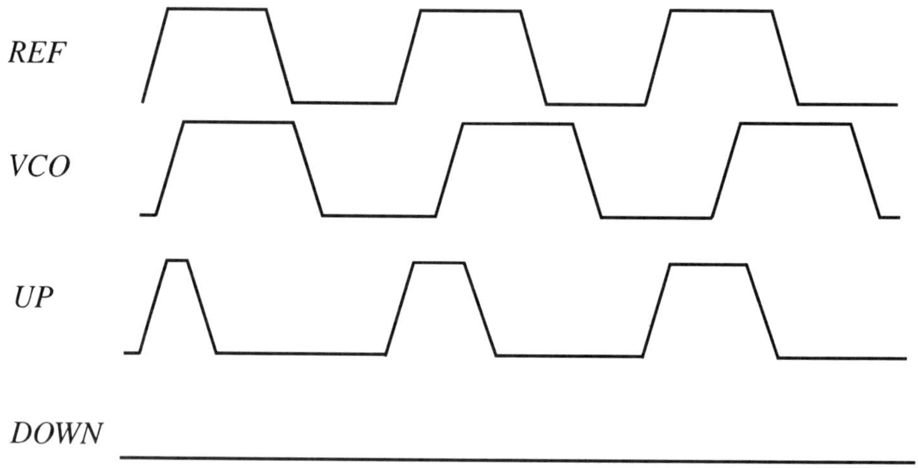

FIGURE E.6 Input and output waveforms for the PFD.

uously as shown in Figure E.6. This will inject charge into the charge pump capacitor, C_P, which in turn results in an increase in the output voltage, V_{out}, to adjust the VCO frequency.

The voltage on the capacitor grows without bound if the input and output do not have the right phase relationship and therefore no static phase error can persist under the locked condition. The PFD/charge-pump architecture has two advantages over the RC lowpass architecture: 1. capture range is limited only by the VCO tuning range; 2. steady-state error is zero.

As long as the dynamics of the loop are much slower than the signal, the charge pump can be treated as a continuous time integrator. Usually a zero is introduced by addition of a resistor in series with the charge pump capacitor to improves loop stability. The phase domain block diagram of Figure E.4 is valid for charge pump PLLs with the following filter transfer function

$$K_P \cdot H(s) = \frac{I}{2\pi C_P} \cdot \frac{\tau_z s + 1}{s} \qquad \text{(E.11)}$$

where I and C_P are the current source and the capacitor in Figure E.5, respectively. Equation (E.11) leads to the following closed-loop transfer function

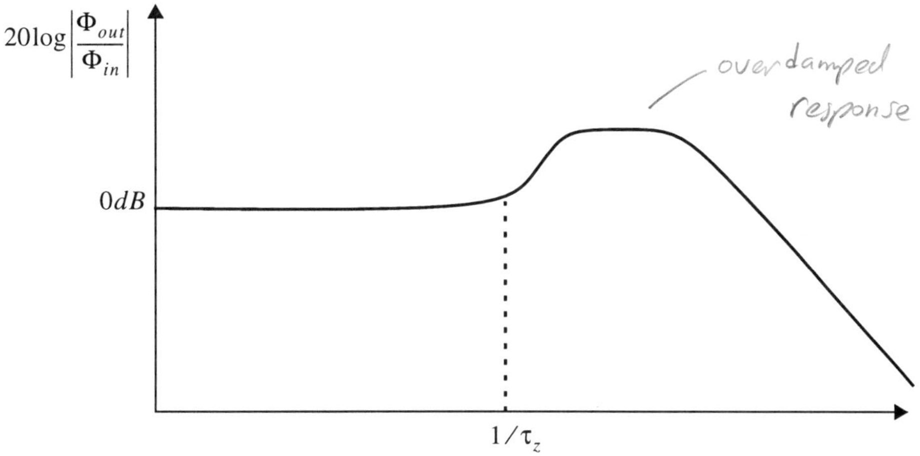

FIGURE E.7 The phase transfer function of the second order charge pump PLL

$$\frac{\Phi_{out}(s)}{\Phi_{in}(s)} = \frac{\tau_z s + 1}{s^2 / \left(K_V \frac{I}{2\pi C_P}\right) + \tau_z s + 1} \qquad \textbf{(E.12)}$$

Figure E.7 shows this transfer function, which results in a zero steady-state phase error for the charge pump PLL.

E.1.3 Frequency Multiplication

In many applications it is desirable to generate a low noise, high frequency signal. As discussed earlier, crystal oscillators provide very stable signal sources; however, crystal oscillators with high resonant frequencies are hard to make. Also it is not possible to change the output frequency of a crystal oscillator by a great amount. For these reasons, frequency synthesizers use a programmable frequency divider in the feedback path of a PLL to generate a stable high frequency and programmable output frequency from a single low frequency signal source [70]-[76]. By introducing a frequency divider in the feedback path of a PLL, frequency multiplication can be achieved as shown in Figure E.8. The division may be performed using an analog frequency divider [77]-[92] or by a digital synchronous or asynchronous counter [93]-[106].

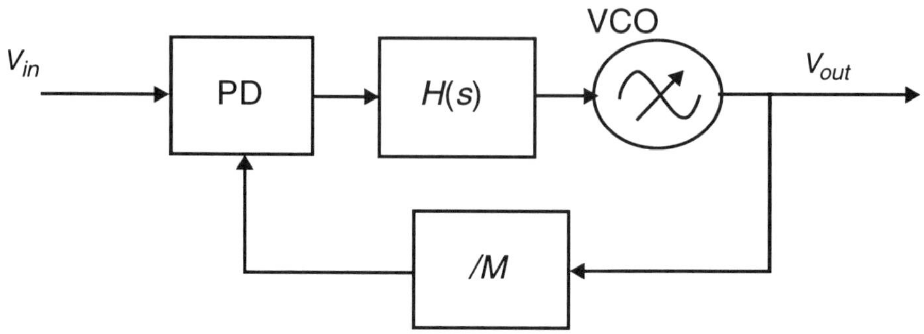

FIGURE E.8 PLL with frequency divider in the feedback path.

The feedback loop adjusts the VCO control voltage in such a way that the frequency of the two inputs to the phase/frequency detector have a constant phase difference and are, therefore, at the same frequency[1]. For this to happen the VCO output frequency has to be M times larger than the input frequency.

An ideal frequency divider divides the phase of the input signal by its division ratio, M, and, hence, in phase domain it is modeled as an attenuation by a factor of $1/M$. The closed-loop transfer function for a charge pump PLL with a divide by M in the feedback path is easily calculated to be

$$\frac{\Phi_{out}(s)}{\Phi_{in}(s)} = \frac{(\tau_z s + 1)}{s^2/K + (\tau_z s + 1)/M} \qquad \textbf{(E.13)}$$

It is instructive to compare this transfer function with that of (E.12), at very low and very high offset frequencies. When $s \to 0$, (E.13) reduces to M, while (E.12) reduces to 1. Consequently, in the case of frequency multiplication, the low frequency input phase variations get multiplied by M. For $s \to \infty$, the transfer functions in both (E.12) and (E.13) similarly reduce to $K\tau_z/s$ and show similar behavior.

1. It is possible for the frequency of the two inputs to the phase detector to have a fractional relationship as discussed in [64], however the loop can be designed in such a way that only locking to the fundamental occurs.

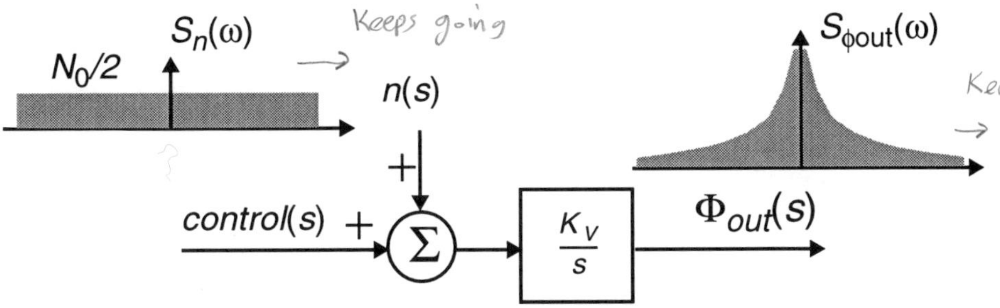

FIGURE E.9 The equivalent model for the VCO.

E.2 Phase Noise and Jitter in PLLs

Having established the essential features of PLLs, the effect of noise in PLLs can be analyzed. A simplified model for the VCO phase noise is applied to first and higher order loops.

E.2.1 VCO Noise

Assuming that phase noise of the VCO is dominated by its phase noise in $1/f^2$ region, the VCO can be modeled as an noiseless VCO which has an additive white noise at the input as shown in Figure E.9, were $n(s)$ is a white noise source, with the double-sideband power spectral density of $N_0/2$. In the absence of variations in the control voltage, the phase output of the VCO to this white noise source will have a power spectrum with the slope of $1/f^2$. This result can be easily understood noting that the VCO acts as an ideal integrator, which is an LTI system, and therefore the output power spectrum can be calculated in terms of the input power spectrum [138][139], namely,

$$S_{\phi_{out}}(\omega) = \left|\frac{K_V}{j\omega}\right|^2 S_n(\omega) = \frac{K_V^2 N_0/2}{\omega^2} \qquad \textbf{(E.14)}$$

Note that $N_0/2$ is chosen in such as way that $S_{\phi out}(\omega)$ corresponds to the phase noise of the VCO in the $1/f^2$ region[1]. This simplified model will be used to predict the behavior of the PLL in presence of noise.

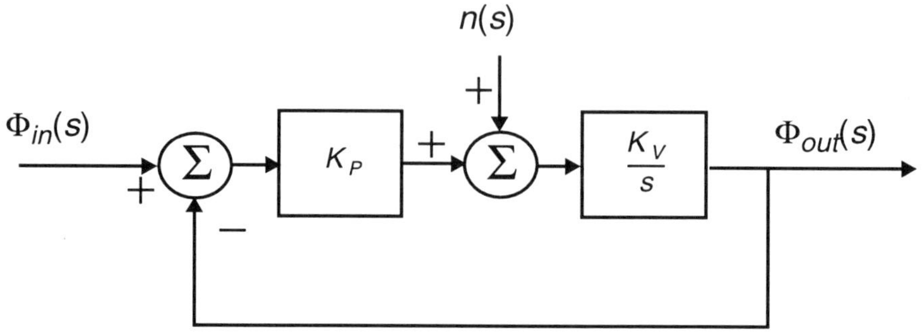

FIGURE E.10 The equivalent first order PLL model with VCO noise.

E.2.2 Other Sources of Noise

Usually phase detectors are not a major source of noise in a PLL. Their noise has been studied to some extent in [67][76][107][108]. The frequency divider in the feedback path may have a significant contribution to the total phase noise of the PLL depending on its implementation and other properties of the loop. Due to their resonant nature, analog regenerative [77]-[90] and injection locked [91][92] dividers are generally less noisy than their digital synchronous and asynchronous counterparts [93]-[106].

E.2.3 Phase Noise and Jitter in First Order Loops

Using the simplified VCO model of Figure E.9, the PLL can be modeled using the equivalent block diagram of Figure E.10. Assuming an ideal phase detector[1], there are two sources of noise which affect the phase noise of the output, Φ_{out}.These two are the VCO phase noise and the phase noise of the input. Assuming that the phase noise of the input is not correlated with the phase noise of the VCO[2], the phase noise power spectrum at the output can be calculated using superposition. In other words, the output spectrum due to each source can be evaluated independently and the total phase noise will be given by their sum.

---

1. The spectrum of the output is related to the spectrum of the phase through a nonlinear phase modulation, and for that reason the spectrum of the output voltage will not grow without bound as does the spectrum of the Φ_{out} does. This is discussed in much more detail in Appendix B and Chapter 5.

1. Effect of phase detector nonideality has been discussed in [64].

2. This is a good assumption unless there is an explicit noise source which dominates the phase noise of both the input and the VCO. Examples of such sources are substrate and supply noise.

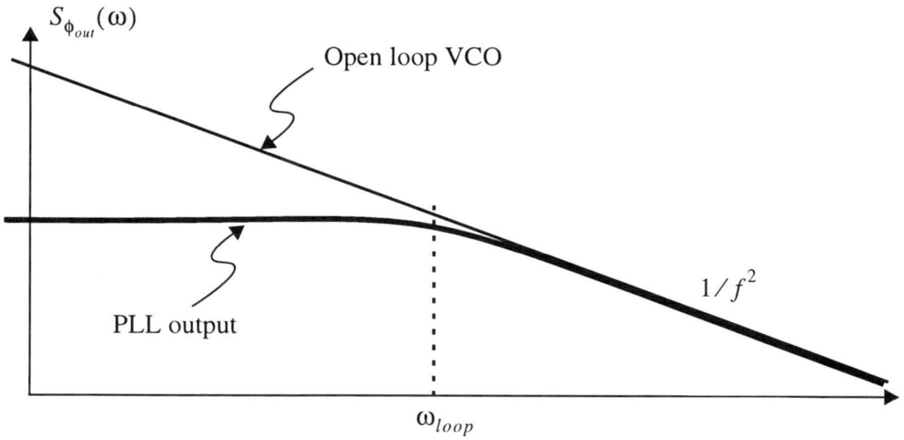

FIGURE E.11 Output phase noise spectrum with an noiseless input.

Assuming an noiseless input, the effect of VCO phase noise can be calculated using the transfer function from $n(s)$ to $\Phi_{out}(s)$, which is

$$\frac{\Phi_{out}(s)}{n(s)} = \frac{K_V}{K_P K_V + s} \tag{E.15}$$

and therefore,

$$S_{\phi_{out}}(\omega) = \frac{N_0}{2}\frac{K_V^2}{(K_P K_V)^2 + \omega^2} \tag{E.16}$$

which is shown in Figure E.11. Comparing (E.14) and (E.16), it is evident that the phase noise of the output is the same as the phase noise of the VCO for offset frequencies larger than ω_{loop}. This should be intuitively clear as the loop adjusts VCO's control voltage to compensate for its slow random variations which are slower than the loop's dynamics. However, it is unable to react fast enough to fast random changes in the VCO output and hence, they appear directly on the output, as can be seen in Figure E.11.

The time domain view of this concept is shown in the timing jitter vs. delay graph of Figure E.12. As mentioned in CHAPTER 2, in an open loop oscillator the timing jitter grows without bound as the delay from the reference edge, τ, increases. However, in a phase locked loop, timing jitter does not increase for time scales larger than loop

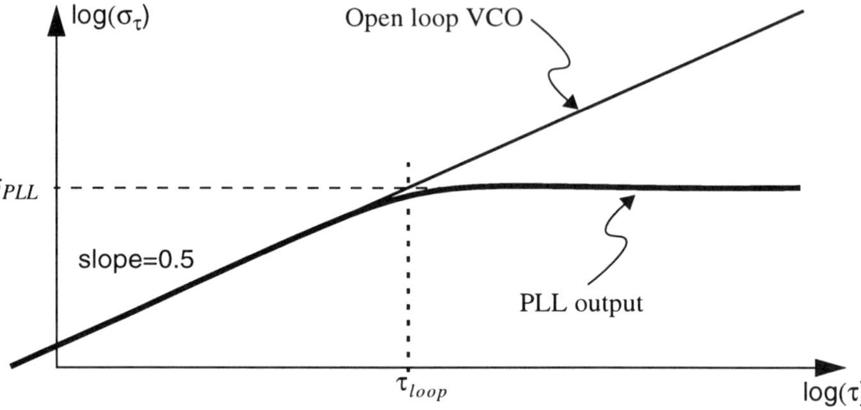

FIGURE E.12 Output timing jitter with an noiseless input.

time constant, τ_{loop}, [48] because the feedback adjusts the VCO control voltage so that the VCO phase follows the input jitter, as shown in Figure E.12. The amount of jitter at the plateau is usually referred to as the PLL jitter and is shown with σ_{PLL} hereafter.

An actual expression for jitter vs. τ in a PLL can be obtained using (A.5). It can be shown that in a first order PLL with a bandwidth of ω_{loop}, the timing jitter is related to τ through

$$\sigma_\tau^2 = \frac{2\pi^2 N K_V^2}{\omega_0^2} \cdot \frac{1}{\omega_{loop}} \cdot \left(1 - e^{-\omega_{loop}\tau}\right) \qquad \textbf{(E.17)}$$

where ω_0 is the center frequency of the output. For $\tau \ll \tau_{loop}$, (E.17) reduces to

$$\sigma_\tau^2 = \frac{2\pi^2 N K_V^2}{\omega_0^2} \cdot \tau \qquad \textbf{(E.18)}$$

as shown in Figure E.12.

Now assume an noiseless VCO and evaluate the response of the loop to the phase variations in the input, Φ_{in}. The input is usually generated by another oscillator, which will have its own phase noise characteristics. Taking into account only the phase noise in the $1/f^2$ region, its power spectrum can be written as $S_{\phi_{in}}(\omega) = \alpha/\omega^2$, where α is

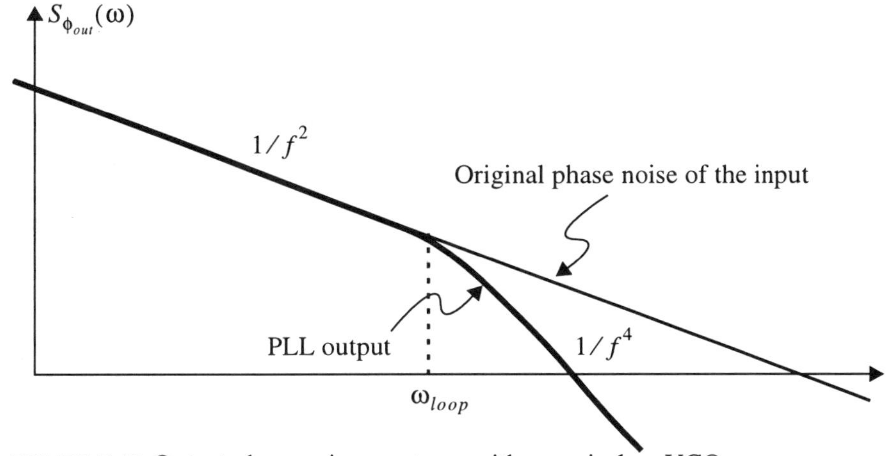

FIGURE E.13 Output phase noise spectrum with an noiseless VCO.

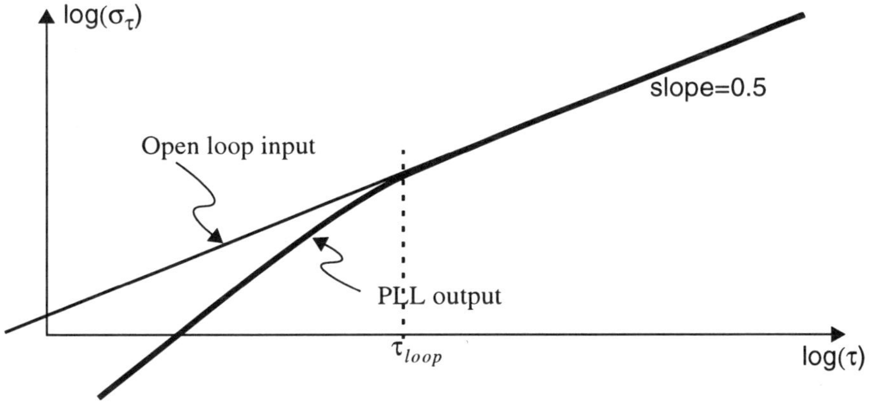

FIGURE E.14 Output timing jitter with an noiseless VCO.

a constant characterizing the phase noise of the input. Using (E.2), the power spectrum of the output can be easily calculated to be

$$S_{\phi_{out}}(\omega) = \frac{\alpha}{\omega^2} \cdot \frac{(K_P K_V)^2}{(K_P K_V)^2 + \omega^2} \tag{E.19}$$

which has the power spectrum shown in Figure E.13. The corresponding time domain picture is shown in Figure E.14.

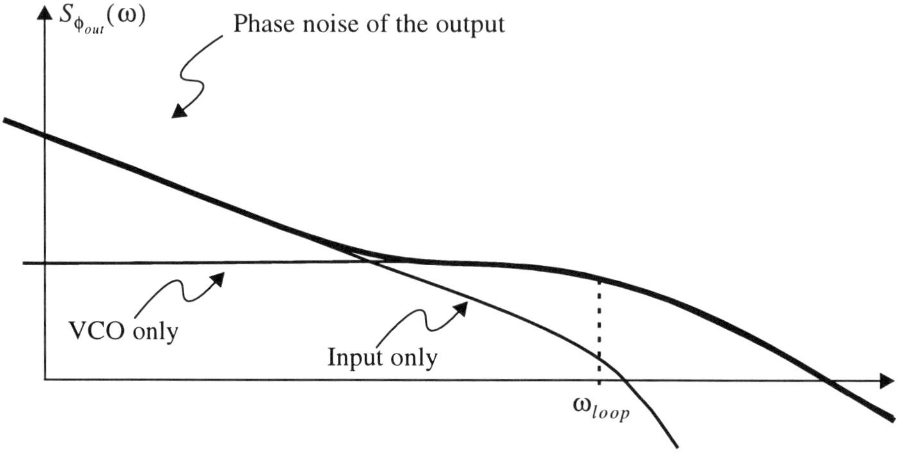

FIGURE E.15 Output phase noise spectrum with a low noise input.

The phase noise of the input can be larger or smaller than the phase noise of the VCO depending on the application in which the PLL is being used. In applications such as microprocessor clock distribution and frequency synthesis, the input usually has a much smaller phase noise than the VCO, and therefore the total effective output phase noise of the PLL will have a shape similar to Figure E.15.

As can be seen from Figure E.15, phase noise is dominated by the input phase noise for small offset frequencies and by the VCO phase noise for large frequency offsets. Phase noise pedestals, such as the one in Figure E.15 are common in synthesizers outputs. Figure E.16 shows timing jitter vs. delay, τ, for this case[1].

In other applications, such as clock recovery, the phase noise of the input signal can be comparable to, or even larger than, the phase noise of the VCO. If that is the case, the phase noise spectrum and timing jitter of the output can have a different shape, as shown in Figure E.17 and Figure E.18

1. Note that if the input signal has better frequency stability compared to the internal time base used in the phase noise/jitter measurement system, phase noise at low offsets (jitter at large delay times) will be dominated by the phase noise (jitter) of the measurement system.

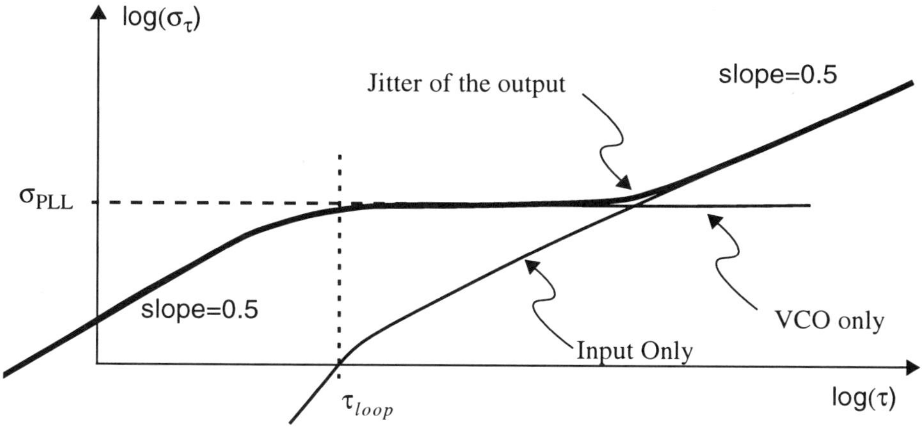

FIGURE E.16 Output timing jitter with a low noise input.

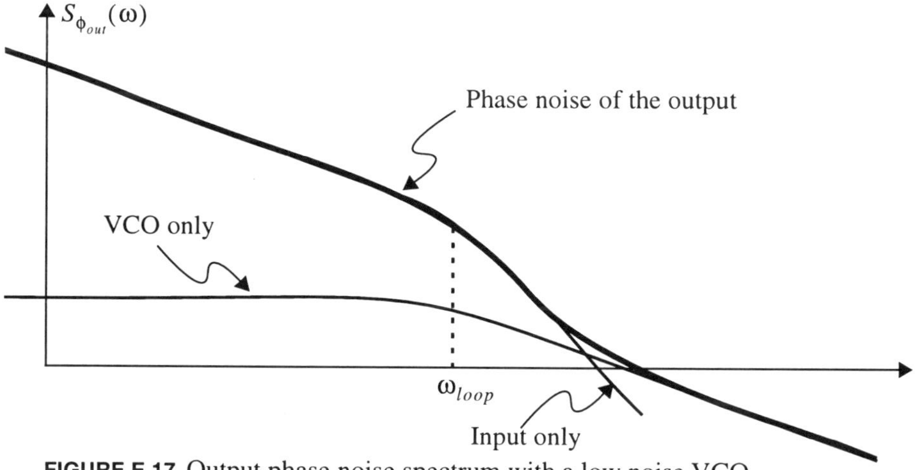

FIGURE E.17 Output phase noise spectrum with a low noise VCO.

E.2.4 Jitter and Phase Noise in Higher Order Loops

Once phase noise behavior in first order loops is understood, it is easy to extend the concept to higher order loops. Consider the example of a charge-pump PLL with a compensation zero described by the phase transfer function (E.12). The transfer function from VCO input noise to output phase is easily calculated to be

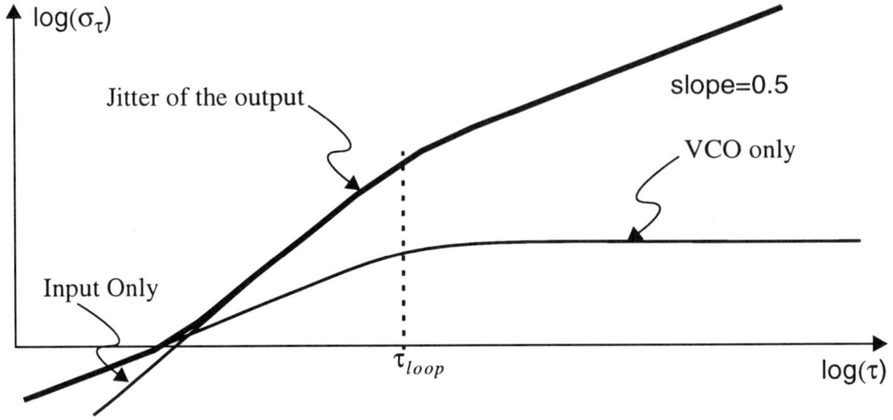

FIGURE E.18 Output timing jitter with a low noise VCO.

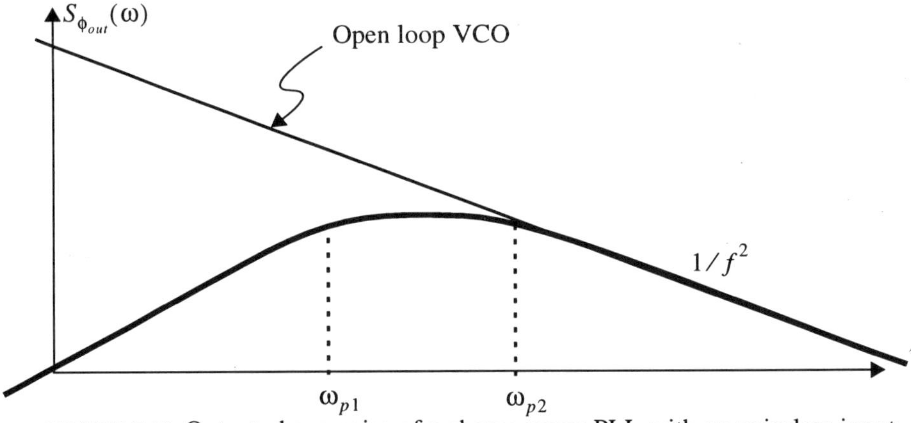

FIGURE E.19 Output phase noise of a charge pump PLL with an noiseless input.

$$\frac{\Phi_{out}(s)}{n(s)} = \frac{2\pi C_P}{I} \cdot \frac{s}{s^2 / \left(K_V \frac{I}{2\pi C_P}\right) + \tau_z s + 1} \tag{E.20}$$

The output phase noise spectrum with a noiseless input signal can be calculated in a fashion similar to (E.16). It will have a spectrum similar to the one shown in Figure E.19. The transfer function for the noise of the input is given by (E.12) and shown in Figure E.7. Therefore, the overall phase noise at the output will have the

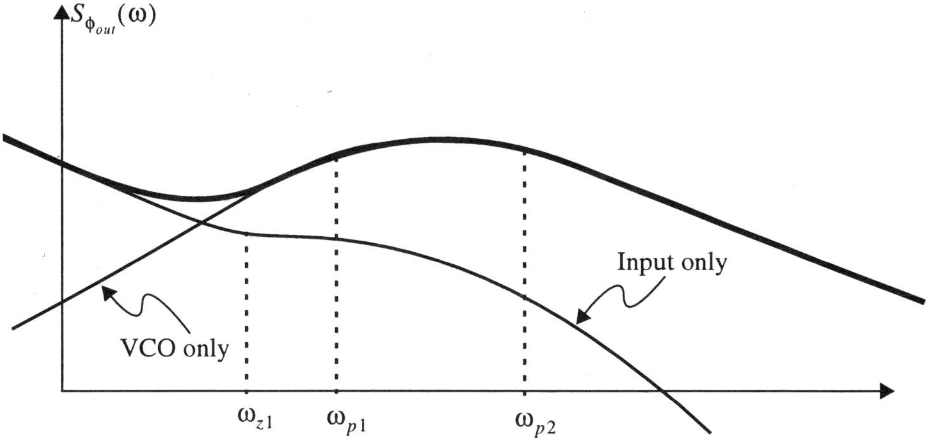

FIGURE E.20 Output phase noise of a charge pump PLL with a low noise input.

form Figure E.20. As can be seen, the phase noise is still dominated by the VCO at large offset frequencies and by the input at small offsets. The bump in the phase noise spectrum of the output is usually referred to as *jitter peaking*.

E.2.5 The Effect of the Frequency Divider

Assuming that the phase detector is not a major source of noise in the PLL, the effect of the frequency divider is to amplify the phase noise of the input by a factor of $20\log(M)$[1]. Any excess phase noise due to divider nonideality will be added to this. Therefore, for a charge pump PLL with a frequency divider in the feedback path, the phase noise spectrum will have a shape similar to Figure E.21.

1. The low frequency input phase variations are multiplied by M at the output. Therefore the phase noise power spectrum of the output at low offset frequencies will be M^2 times the input phase fluctuations and hence $20\log(M)$.

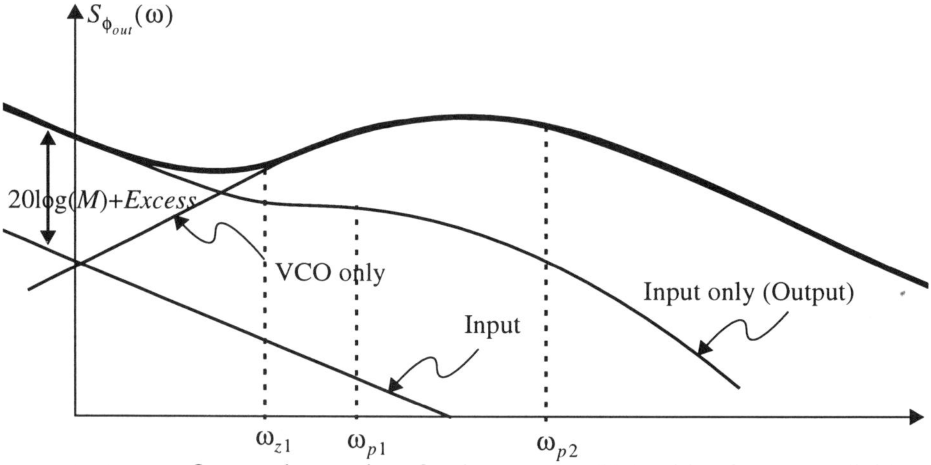

FIGURE E.21 Output phase noise of a charge pump PLL with a frequency divider.

APPENDIX F Describing Function Analysis of Oscillators

The effect of the nonlinearity on the oscillator amplitude can be evaluated using describing function analysis [4]-[12]. This appendix presents a simplified, but general, approach to amplitude prediction using describing functions.

Consider the forward path transconductance block, G, in the two-port model of Figure F.1 It will be assumed that it consists of a memoryless nonlinearity as shown in Figure F.2. In an oscillator with high tank Q, the output voltage of the frequency selective network of Figure F.1 will be very close to a sinusoidal voltage even for a periodic non-sinusoidal input current, as shown in Figure F.3.

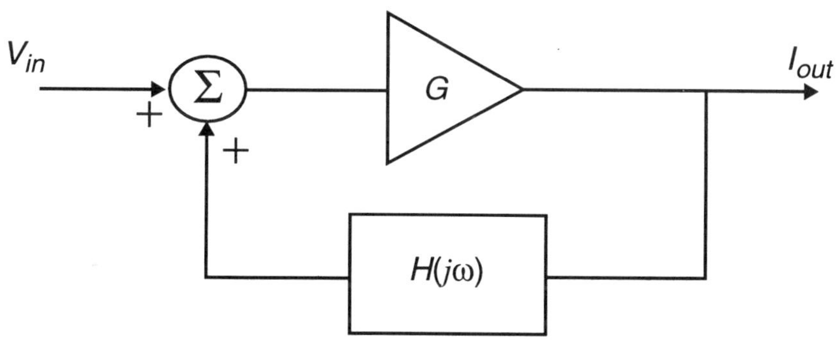

FIGURE F.1 Simplified two-port model of an oscillator.

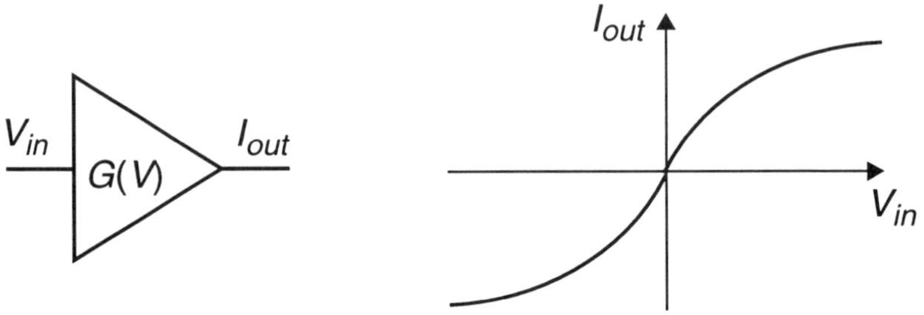

FIGURE F.2 The memoryless nonlinearity in the forward path.

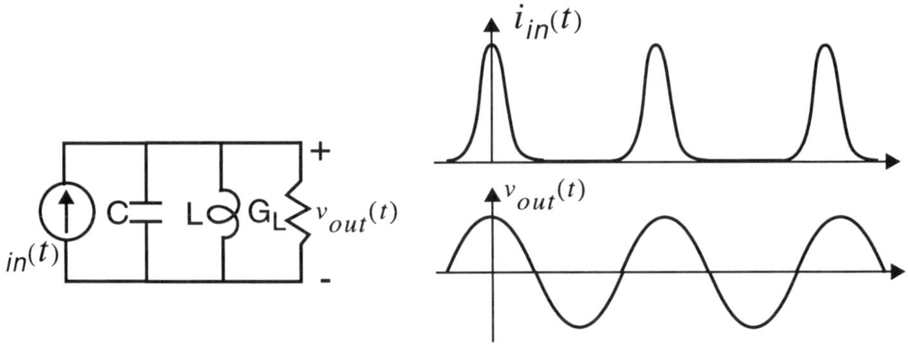

FIGURE F.3 The input current and output voltage for a high Q tank.

Since the output voltage of the frequency selective network is the input to the nonlinear transconductance block, the response of the nonlinear block, G, to a sinusoidal input should be characterized. Although the output current of the nonlinear transconductance will not be sinusoidal, the frequency selective network will mainly pass the fundamental term of the input since it will attenuate all the other harmonics significantly. Therefore, it is the gain from the input sinusoidal voltage to the fundamental component of the output current that determines the loop gain.

Based on the foregoing observations, the nonlinear transconductance is assumed to be driven with a sinusoidal input of amplitude V_1. In the most general case, the output will have all the terms of the Fourier series. Thus, for an input voltage of the following form,

$$v_{in}(t) = V_1 \cos(\omega_0 t) \qquad \textbf{(F.1)}$$

the output current can be written as[1]

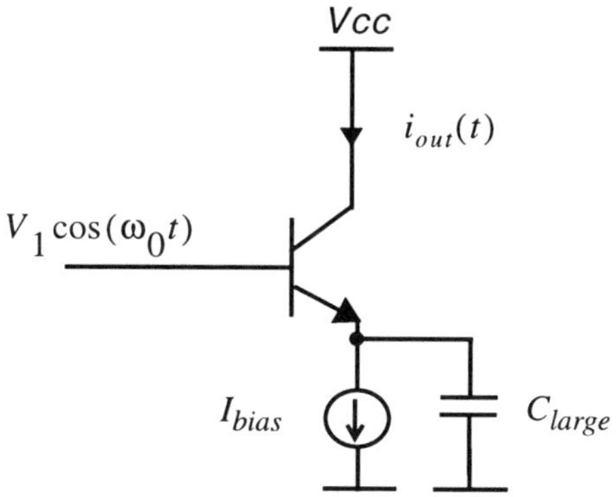

FIGURE F.4 A bipolar transistor with a sinusoidal current drive.

$$i_{out}(t) = \sum_{n=1}^{\infty} I_n \cos(n\omega_0 t) \quad \textbf{(F.2)}$$

The amplitude ratio of the fundamental output component to the input is the magnitude of the describing function, which will be denoted as $G_m(V_1)$, or G_m for short. Thus,

$$G_m(V_1) = \frac{I_1}{V_1} \quad \textbf{(F.3)}$$

This naming convention underscores that G_m is the effective large signal transconductance of the nonlinear block at ω_0.

As an example, consider the case of an ideal bipolar transistor biased with a current source in parallel with a large capacitor, as shown in Figure F.4. Assuming ideal bipolar transistor behavior, the collector current is related to the input voltage through an exponential relation,

$$I_C = I_S \exp\left(\frac{V_{BE}}{V_T}\right) \quad \textbf{(F.4)}$$

1. There is no constant phase in the cosines of (F.2) because the nonlinearity considered here is memoryless. However, the active gain block will contribute some phase. This phase has an effect on the phase noise as discussed in CHAPTER 3.

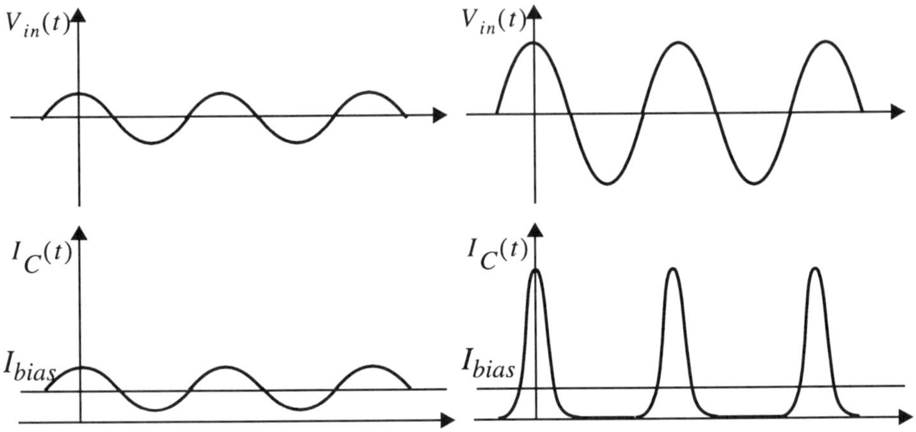

FIGURE F.5 The input voltage and output current for a) small sinusoidal input, b) large sinusoidal input

where I_S is the reverse saturation current of the base emitter junction, V_{BE} is the base emitter voltage and V_T is the thermal voltage, given by

$$V_T = \frac{kT}{q} \tag{F.5}$$

in which k is Boltzmann's constant, q is the electron charge and T is the temperature in kelvins. The output will be sinusoidal for small values of V_1, but becomes more impulsive as the input amplitude grows, as shown in Figure F.5. This impulsive behavior affects the properties of the noise sources in the circuit and has an important effect on the phase noise of the oscillators, as discussed in CHAPTER 4.

Although it is possible to derive the large signal transconductance, G_m, for the bipolar transistor in terms of modified Bessel functions [10], investigating the two extreme cases of very large and very small values of V_1 provides important information.

For very small values of V_1, the small signal assumption holds and the output grows linearly with input. Therefore,

$$G_m(V_1) = g_m = \frac{I_{tail}}{V_T} \qquad V_1 \ll V_T \tag{F.6}$$

where g_m is the small signal transconductance of the transistor.

For large input amplitude, the output current will consist of sharp spikes of current, whose average value necessarily equals I_{bias}. Therefore, the fundamental component of the output current can be approximated by [69]

$$I_1 = \frac{2}{T}\int_0^T i(t)\cos(\omega_0 t)dt \approx \frac{2}{T}\int_0^T i(t)dt = 2I_{bias} \tag{F.7}$$

where T is the period of the oscillation. For large values of V_1, the spikes will be very thin and tall and will occur at the peak of the cosine function. The approximation in (F.7) holds as long as the spikes are sharp enough so that the cosine can be approximated as 1 for the duration of the spike. Using (F.3) and (F.7), the describing function for large values of V_1 can be written as

$$G_m(V_1) = \frac{2I_{tail}}{V_1} \qquad V_1 \gg V_T \tag{F.8}$$

As can be seen, the large signal transconductance is inversely proportional to the input voltage amplitude for large values of input voltage. This inverse proportionality provides a negative feedback mechanism that stabilizes the amplitude of oscillations by reducing the effective gain as the amplitude grows.

To gain more insight, the value of G_m from (F.6) and (F.8), together with the actual value obtained from the complete analysis [10], is plotted versus the input amplitude, V_1 in Figure F.6. As can be seen, (F.8) gives a good approximation for the large signal transconductance of a bipolar transistor as long as $V_1 \gg V_T$.

It is noteworthy that (F.8) is valid for other types of devices with monotonic nonlinearity, such as MOS transistors, vacuum tubes, etc. as long as V_1 is larger than a characteristic voltage that depends on the particular device of interest. This universality holds because the only assumption used to obtain (F.8) is that the spikes are so thin that the cosine function can be approximated as 1 for the duration of the spike.

Describing function analysis can be applied to calculate the amplitude and frequency of oscillation. As an example, consider the common drain MOS Colpitts oscillator of Figure F.7.

The large signal equivalent circuit for the oscillator of Figure F.7 is shown in Figure F.8. The tank voltage amplitude is related to V_1 through

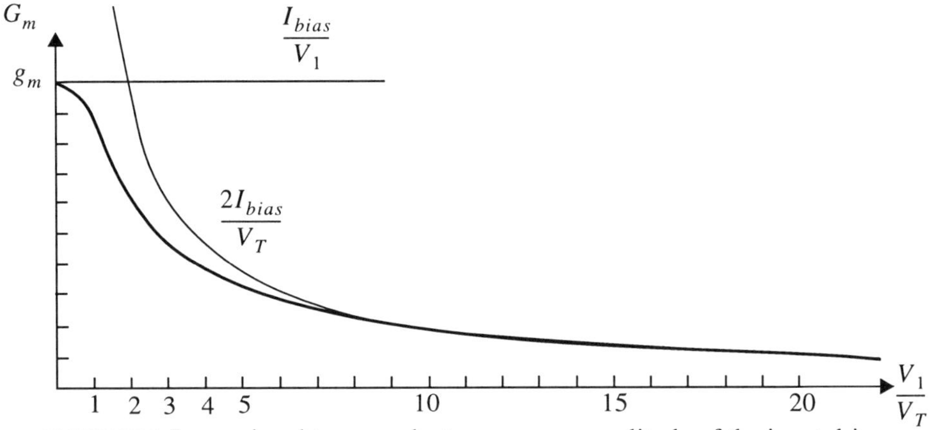

FIGURE F.6 Large signal transconductance versus amplitude of the input drive.

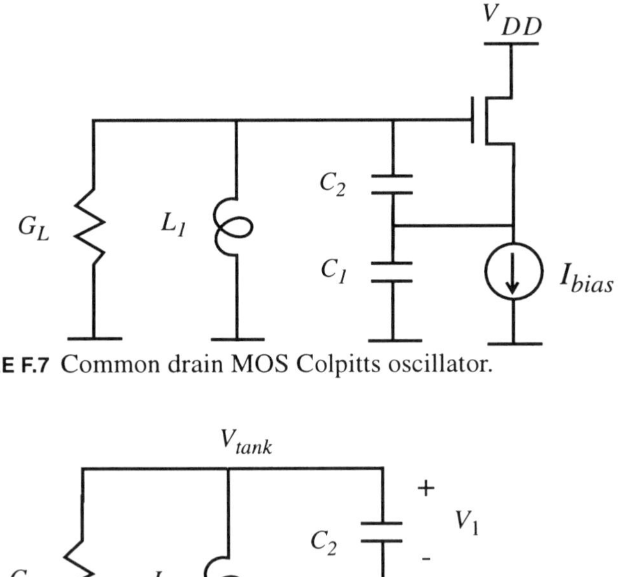

FIGURE F.7 Common drain MOS Colpitts oscillator.

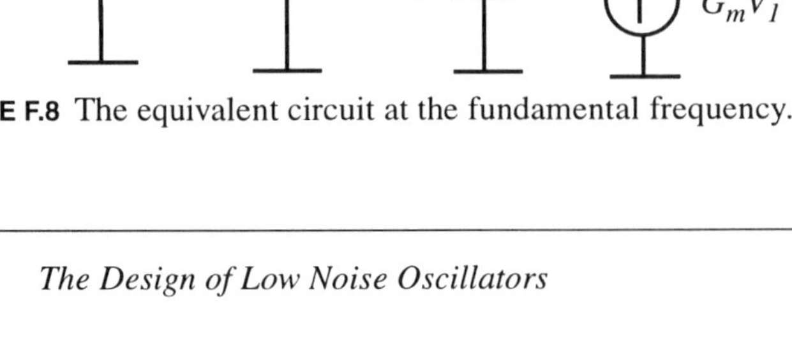

FIGURE F.8 The equivalent circuit at the fundamental frequency.

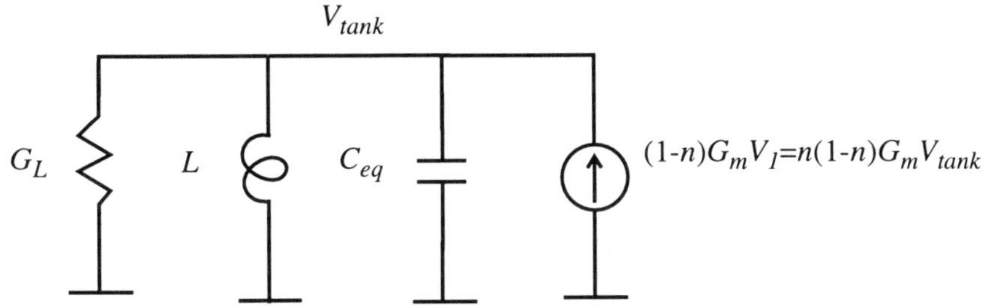

FIGURE F.9 The simplified large signal equivalent circuit

$$V_1 = V_{tank}\frac{C_1}{C_1 + C_2} = nV_{tank} \tag{F.9}$$

where

$$n = \frac{C_1}{C_1 + C_2} \tag{F.10}$$

is the capacitive voltage division ratio. The capacitive divider scales the current by a ratio of 1-n. Therefore, the equivalent circuit reduces to that of Figure F.9, in which C_{eq} is the series combination of C_1 and C_2.

In steady-state, tank current is related to the tank voltage through

$$I_{tank} = Y_{tank}V_{tank} = \left(G_L + jC_{eq}\omega + \frac{1}{jL\omega}\right)V_{tank} = n(1-n)G_mV_{tank} \tag{F.11}$$

where Y_{tank} and G_L are the admittance and effective parallel conductance of the tank, respectively. For (F.11) to hold, we should have

$$\omega_0 = \frac{1}{\sqrt{LC}} \tag{F.12}$$

and

$$G_L = n(1-n)G_m \tag{F.13}$$

Using (F.8) and (F.9), the tank voltage amplitude is calculated to be

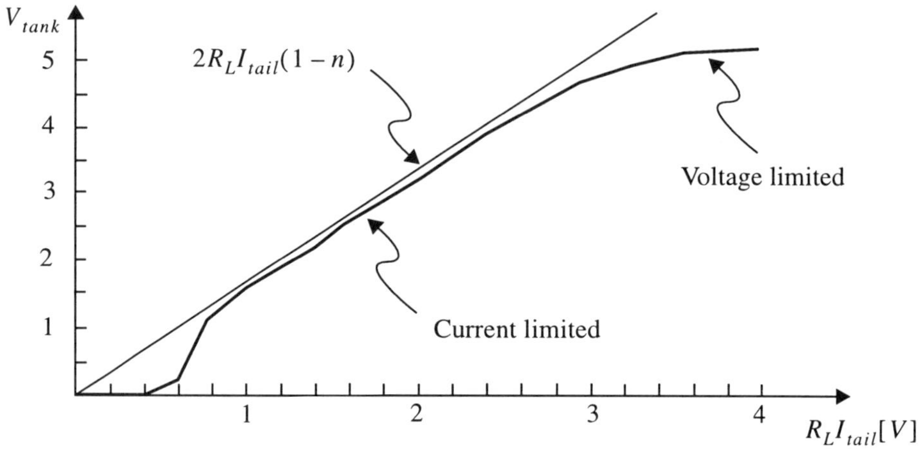

FIGURE F.10 Tank amplitude vs. R_LI_{tail} for common drain Colpitts oscillator.

$$V_{tank} = 2R_LI_{tail}(1-n) \qquad \text{(F.14)}$$

As can be seen from (F.14), for small C_1/C_2 ratios, the tank voltage amplitude is about twice the product of tail current and effective tank resistance. This mode of operation is usually referred to as *current limited.*

Note that (F.14) breaks down for small values of I_{tail} in accordance with (F.6). It also fails for *large* values of I_{tail} as V_{tank} approaches V_{DD}. This failure happens as the MOS transistor enters the ohmic region (or saturation for a bipolar transistor) for part of the period, therefore violating the assumptions leading to (F.14). The value of V_{tank} for which this happens depends on the supply voltage, and therefore this regime of operation is known as *voltage limited.*

Figure F.10 shows the simulated tank amplitude for the common drain Colpitts oscillator of Figure F.7 versus R_LI_{tail}, with $C1/C2$=0.2. As can be seen (F.14) is accurate in the current limited mode, but loses its validity in the voltage limited regime.

Bibliography

[1] L. L. Gossack, "Cellular Radio Telephone Systems," *U.S. Industrial Outlook 1992: Radio Communications and Detection Equipment*, publisher: U.S. Department of Commerce, Washington D.C., 1992.

[2] R. Schneiderman, "For Offshore Wireless Markets, the Accent is on Rapid Growth," *Microwave and RF*, Aug. 1994.

[3] K. Feher, *Wireless Digital Communications, Modulation & Spread Spectrum Applications*, Prentice Hall, 1995.

[4] H. Poincare, *Les Méthodes Nouvelles de la Mécanique Céleste*, Gauthier-Villars, Paris 1892.

[5] A. Liapounov, "Problème Général de la Stabilité du Mouvement," *Ann. Fac. Sci.*, Toulouse 1907.

[6] B. van der Pol, "The Nonlinear Theory of Electric Oscillations," *Proc. IRE*, vol 22, pp. 1051-1086, Sept. 1934.

[7] N. Minorsky, *Nonlinear Oscillations*, D. Van Nostrand Company, Inc., Princeton,1962.

[8] J. K. Hale, *Oscillations in Nonlinear Systems*, Dover Publications, Inc., New York 1963.

[9] A. A. Andronov, A. A. Vitt and S. E. Khaikin, *Theory of Oscillators*, Pergamon Press Ltd., 1966.

[10] K. K. Clarke and D. T. Hess, *Communication Circuits: Analysis and Design*, Addison-Wesley Publishing Company Inc., 1971.

[11] B. Parzen and A. Ballato, *Design of Crystal and Other Harmonic Oscillators*, John Wiley and Sons, New York, 1983.

[12] P. A. Cook, *Nonlinear Dynamical Systems,* Prentice Hall, New York 1994.

[13] L. M. Hull and J. K. Clapp, "A Convenient Method for Referring Secondary Frequency Standards to a Standard Time Interval," *Proc. IRE*, vol. 17, Feb. 1929.

[14] J. L. Stewart, "Frequency Modulation Noise in Oscillators," *Proc. IRE*, pp. 372-6, March 1956.

[15] W. A. Edson, "Noise in Oscillators," *Proc. IRE*, pp. 1454-66, Aug. 1960.

[16] J. A. Mullen, "Background Noise in Nonlinear Oscillators," *Proc. IRE*, pp. 1467-73, Aug. 1960.

[17] R. Grivet and A. Blaquiere, "Nonlinear Effects of Noise in electronic Clocks," *Proc. IRE*, pp. 1606-14, Nov. 1963.

[18] E. J. Baghdady, R. N. Lincoln and B. D. Nelin, "Short-Term Frequency Stability: Characterization, Theory, and Measurement" *Proc. IEEE,* vol. 53, pp. 704-722, July 1965.

[19] E. Hafner, "The Effect of Noise in Oscillators," *Proc. IEEE*, vol. 54, pp. 179-198, Feb. 1966.

[20] L. S. Cutler and C. L. Searle, "Some Aspects of the Theory and Measurement of Frequency Fluctuations in Frequency Standards," *Proc. IEEE,* vol. 54, pp. 136-154, Feb. 1966.

[21] M. Lax, "Classical Noise V. Noise in Self-Sustained Oscillators," Phys. Rev., CAS-160, vol. 160, no. 2, pp. 290-307, Aug. 1967.

[22] D. B. Leeson, "A Simple Model of Feedback Oscillator Noise Spectrum," *Proc. IEEE,* vol. 54, pp. 329-330, Feb. 1966.

[23] J. A. Barnes, *et al*, "Characterization of Frequency Stability," *IEEE Trans. Instr. & Meas.*, vol. IM-20, no. 2, pp. 105-20, May 1971.

[24] J. Rutman, "Characterization of Phase and Frequency Instabilities in Precision Frequency Sources; Fifteen Years of Progress," *Proc. IEEE,* vol. 66, pp. 1048-1174, Sept. 1978.

[25] W. P. Robins, *Phase Noise in Signal Sources*, Peter Peregrinus Ltd., London 1982.

[26] A. A. Abidi and R. G. Meyer, "Noise in Relaxation Oscillators," *IEEE J. Solid-State Circuits,* vol. SC-18, pp. 794-802, Dec. 1983.

[27] V. C. Vannicola and P. K. Varshney, "Spectral Dispersion of Modulated Signals due to Oscillator Phase Instability: White and Random Walk Phase Model" *IEEE Trans. Comm.*, vol. COM-31, no.7, pp. 886-95, July 1983.

[28] H. B. Chen, A. van der Ziel and K. Amberiadis, "Oscillator with Odd-Symmetrical Characteristics Eliminates Low-Frequency Noise Sidebands," *IEEE Trans. on Circ. and Syst.* vol. CAS-31, NO. 9, Sept. 1984.

[29] C. A. M. Boon, I. W. J. M. Rutten and E. H. Nordholt, "Modeling the Phase Noise of RC Multivibrators," *Proc. 27th Midwest Symp. Circuits and Systems*, vol. 2, pp. 421-4, June 1984.

[30] G.J. Foschini and G. Vannucci, "Characterizing Filtered Light Waves Corrupted by Phase Noise," *IEEE Trans. Info. Theory*, vol.34, no.6, pp. 1437-48, Nov. 1988.

[31] F. X. Kaertner, "Determination of the Correlation Spectrum of Oscillators with Low Noise," *IEEE Trans. Microwave Theory and Techniques*, vol. 37, no. 1, January 1989.

[32] F. X. Kaertner, "Analysis of white and $f^{-\alpha}$ Noise in Oscillators," *International Journal of Circuit Theory and Applications*, vol. 18, pp. 485-519, 1990.

[33] V. Rizzoli, A. Costanzo, F. Mastri and C. Cecchetti, "Harmonic-Balance Optimization of Microwave Oscillators for Electrical Performance, Steady-State Stability, and Near-Carrier Phase Noise" *Proceedings of IEEE/MTT-S,* pp. 1401-4 vol.3, May 1994.

[34] V. Rizzoli, F. Mastri and D. Masotti, "General Noise Analysis of Nonlinear Microwave Circuits by the Piecewise Harmonic-Balanced Technique," *IEEE Trans. on MTT*, vol. 42, no. 5, pp. 807-819, May 1994.

[35] A. Demir, A. L. Sangiovanni-Vincentelli, "Simulation and Modeling of Phase Noise in Open-Loop Oscillators", *Proc. of CICC,* 1994.

[36] M. N. Tutt, D. Pavlidis, A. Khatibzadeh and B. Bayraktaroglu, "The Role of Baseband Noise and its Upconversion in HBT Oscillator Phase Noise," *IEEE Trans. on Microwave Theory and Techniques*, vol. 43, no. 7, July 1995.

[37] J. Craninckx and M. Steyaert, "Low-noise Voltage Controlled Oscillators Using Enhanced LC-Tanks," *IEEE Trans. Circuits Syst.-II,* vol. 42, pp. 794-904, Dec. 1995.

[38] R. Telichevesky, K. Kundert, I. Elfadel and J. White, "Fast simulation algorithms for RF circuits," *Proceedings of CICC*, pp. 437-44, May 1996.

[39] L. Kazovsky, S. Benedetto and A. Willner, *Optical Fiber Communication Systems*, Artech House, Boston 1996.

[40] Y. Yamamoto, *Fundamentals of Noise Processes*, Stanford University, Class Notes, 1996.

[41] S. Brigati, F. Francesconi and F. Maloberti, "A 200-MHz Crystal Oscillator with 0.5 ppm Long Term Stability," *Proc. 3rd Int. Conf. on Electronics Circ. & Systs.*, vol. 1, pp. 164-7, Oct. 1996.

[42] M. Okumura, and H. Tanimoto, "A Time-Domain Method for Numerical Noise Analysis of Oscillators," *Proceedings of ASP-DAC '97*, pp. 477-482, Jan. 1997.

[43] A. Hajimiri and T. H. Lee, "A General Theory of Phase Noise in Electrical Oscillators," *IEEE J. Solid-State Circuits.* vol. 33, no. 2, Feb. 1998.

[44] J. G. Maneatis and M. A. Horowitz, "Precise Delay Generation Using Coupled Oscillators," *IEEE J. Solid-State Circuits*, vol.28, no.12, pp. 1273-82, Dec. 1993.

[45] J. McNeil, "Jitter in Ring Oscillators," *Proc. ISCAS 94,* June 1994.

[46] T. C. Weigandt, B. Kim and P. R. Gray, "Analysis of Timing Jitter in CMOS Ring Oscillators," *Proc. ISCAS,* June 1994.

[47] B. Razavi, "A Study of Phase Noise in CMOS Oscillators," *IEEE J. Solid-State Circuits,* vol. 31, no. 3, March 1996.

[48] J. McNeill, "Jitter in Ring Oscillators," *IEEE J. Solid-State Circuits,* vol. 32, no. 6, pp. 870-879, June 1997.

[49] A. Hajimiri, S. Limotyrakis and T. H. Lee, "Phase Noise in Multi-Gigahertz CMOS Ring Oscillators," *Proc. Custom Integrated Circuits Conference*, May 1998.

[50] F. Herzel and B. Razavi, "Oscillator Jitter Due to Supply and Substrate Noise," *Proc. Custom Integrated Circuits Conference*, May 1998.

[51] L. Dauphinee, M. Copeland and P. Schvan, "A Balanced 1.5GHz Voltage Controlled Oscillator with an Integrated LC Resonator," *ISSCC Dig. Tech. Papers*, pp.390-1, Feb. 1997.

[52] J. Craninckx, M. Steyaert and H. Miyakawa, "A Fully Integrated Spiral-LC CMOS VCO Set with Prescaler for GSM and DCS-1800 Systems" *Proc. CICC*, pp. 403-6, May 1997.

[53] J. Craninckx and M. Steyaert, "A 1.8-GHz Low-Phase-Noise CMOS VCO Using Optimized Hollow Spiral Inductors" *IEEE J. Solid-State Circ*uits, vol.32, no.5, p. 736-44, May 1997.

[54] T. I. Ahrens, A. Hajimiri and T. H. Lee, "A 1.6GHz, 0.5mW CMOS LC Low Phase Noise VCO Using Bond Wire Inductance," *Proc. of the 1st International Workshop on Design of the Mixed-Mode Integrated Circuits and Applications,* Cancun, July 1997.

[55] A. Hajimiri and T.H. Lee, "Phase Noise in CMOS Differential LC Oscillators," *Proc. VLSI Circuits*, June 1998.

[56] T. I. Ahrens and T. H. Lee, "A 1.4GHz 3mW CMOS LC Low Phase Noise VCO Using Tapped Bond Wire Inductance," *Proc. ISLPED*, pp.16-9, Aug. 1998.

[57] J. H. Shoaf, D. Halford and A. S. Risley, "Frequency Stability Specification and Measurement: High Frequency and Microwave Signals," *N.B.S. Tech. Note 632*, Jan. 1973.

[58] D. Howe, "Frequency Domain Stability Measurements: A Tutorial Introduction," *N.B.S. Tech. Note 679*, National Bureau of Standards, Boulder, CO, 1976.

[59] M. Fischer, "Frequency Stability Measurement Procedures," *Proc. 8th Annual Precision Time and Time Interval Meeting*, Dec. 1976.

[60] A. L. Lance, W. D. Seal, F. G. Mendoza and N. W. Hudson, "Automatic Phase Noise Measurements," *Microwave J.*, vol.20, no.6, pp. 87-90, June 1977.

[61] M. McNamee, "Automate for Improved Phase-Noise Measurement," *Microwaves*, vol.48, no.5, pp. 80-1, May 1979.

[62] D. Scherer, "The 'Art' of Phase Noise Measurement," *Hewlett-Packard RF & Microwave Meas. Symp.*, May 1983.

[63] R. Temple, "Choosing a Phase Noise Measurement Technique-Concept and Implementation," *Hewlett-Packard RF & Microwave Meas. Symp.*, Feb. 1983.

[64] F. M. Gardner, *Phaselock Techniques, 2nd Edition*, John Wiley & Sons, New York, 1979.

[65] U. L. Rohde, J. C. Whitaker and T. T. N. Bucher, *Communications Receiver: Principles and Design, 2nd Edition*, McGraw-Hill, 1997.

[66] W. F. Egan, *Frequency Synthesis by Phase Lock,* John Wiley and Sons 1981.

[67] J, A. Crawford, *Frequency Synthesizer Design Handbook*, Artech House, Inc. 1994.

[68] B. Razavi, *Monolithic Phase-Locked Loops and Clock Recovery Circuits, Theory and Design*, IEEE Press, 1996.

[69] T. H. Lee, *The Design of CMOS Radio-Frequency Integrated Circuits*, Cambridge University Press, Cambridge, 1998.

[70] J. Noordanus, "Frequency Synthesizers - A Survey of Techniques," *IEEE Trans. on Comm.*, vol. COM-18, pp. 632-7, Oct. 1970.

[71] J. Tierney, C. M. Rader, B. Gold, "A Digital Frequency Synthesizer," *IEEE Trans. on Audio and Electroacoustics*, vol. AU-19, no.1, p. 48-57, March 1971.

[72] V. F. Kroupa, *Frequency Synthesis*, Wiley, New York, 1973.

[73] G. Gorski-Popiel (Editor), *Frequency Synthesis: Applications and Techniques*, IEEE Press, New York, 1975.

[74] Manassewitsch, *Frequency Synthesizers*, Wiley, New York, 1976.

[75] W. F. Egan, "The Effect of Small Contaminating Signals in Nonlinear Elements Used in Frequency Synthesis and Conversion," *Proc. IEEE*, vol. 69, no. 7, pp. 797-811, July 1981.

[76] V. F. Kroupa, "Noise Properties of PLL Systems," *IEEE Trans. Comm.*, vol. COM-30, no. 10, pp. 2244-52, Oct. 1982.

[77] J. Groszkowski, "Frequency Division," *Proc. IRE*, vol. 18, no. 11, pp. 1960-70, Nov. 1930.

[78] H. Sterky, "Frequency Multiplication and Division," *Proc. IRE*, vol. 25, no. 9, pp. 1153-73, Sept. 1937.

[79] R. L. Miller, "Fractional-Frequency Generators Utilizing Regenerative Modulation," *Proc. IRE*, vol. 27, no. 7, pp. 446-57, July 1939.

[80] K. K. Agarwal and W. J. Thompson, "Phase Noise Behavior of Frequency Dividers Implemented with GaAs Heterojunction Bipolar Transistors," *Electronics Lett.*, vol. 21, no. 22, pp1005-6, Oct. 1985.

[81] D. Huffman, "Extremely Low Noise Frequency Dividers," *Microwave J.*, pp. 209-10, Nov. 1985.

[82] T. Ohira, T. Hiraoka and H. Kato, "MMIC 14-GHz VCO and Miller Frequency Divider for Low-Noise Local Oscillators," *IEEE Trans. Microwave Theory Techniques*, vol. MTT-35, no. 7, pp. 657-62, July 1987.

[83] M. M. Driscoll, "Phase Noise Performance of Analog Frequency Dividers," *IEEE Trans. Ultrasonic, Ferroelectrics and Freq. Cont.*, vol. 37, no. 4, pp. 295-304, July 1990.

[84] J. Groslambert, M. Olivier and E. Rubiola, "High Spectral Purity Frequency Sources Using Low Noise Regenerative Frequency Dividers," *Proc. 45th Annual Symposium on Freq. Cont.*, pp. 636-9, May 1991.

[85] E. Rubiola, M. Olivier and J. Groslambert, "Phase Noise in the Regenerative Frequency Dividers," *IEEE Tans. Instrumentation and Meas.*, vol. 41, no. 3, June 1992.

[86] G. R. Sloan, "The Modeling, Analysis, and Design of Filter-Based Parametric Frequency Dividers," *IEEE Trans. on MTT*, vol. 41, no. 2, pp. 224-8, Feb. 1993.

[87] O. Llopis, H. Amine, M. Gayral, J. Graffeuil and J. F. Sautereau, "Analytical Model of Noise in an Analog Frequency Divider," *IEEE MTT Int. Microwave Symp. Dig.*, pp. 1033-6, 1993.

[88] H. Amine, O. Llopis, R. Plana, J Graffeuil and J.F. Sautereau, "Conversion Gain and Noise in Microwave Analog Frequency Dividers Using Various Types of FET," *Proc. 23rd European Microwave Conf.*, pp. 774-6, Sept. 1993.

[89] H. Brauns and W. Konrath, "Ultra Low Phase Noise Parametric Frequency Divider for Highest Performance Microwave- and Millimeter-Wave Frequency Sources," *Proc. 25th European Microwave Conf.*, pp. 1155-8, Sept. 1995.

[90] B. Branger, E. LaPorte, J.C. Nallatamby, M. Prigent and L. LaPierre, "Comparison between Simulations and Measurements of Large Signal and Nonlinear noise Behaviors of MMIC Analog Frequency Divider by Two," *IEEE Microwave and Guided Wave Letters*, vol. 7, no. 5, May 1997.

[91] X. Zhang, X. Zhou and A. S. Daryoush, "A Theoretical and Experimental Study of the Noise Behavior of Subharmonically Injection Locked Local Oscillators," *IEEE Trans. on MTT*, vol. 40, no. 5, pp. 895-902, May 1992.

[92] H. Rategh and T. H. Lee, "Superhamonic Injection Locked Oscillators as Low Power Frequency Dividers," *Symp. on VLSI Circ. Dig.*, pp. 132-5, June 1998.

[93] A. A. L'vovich, "Design of Noise-Immune Counter-Type Frequency Dividers," *Elektrosvyaz, Translated in: Telecomm. and Radio Eng.*, part 1, vol. 29, no. 2, pp. 52-5, feb. 1975.

[94] D. Scherer, "Design principles and Measurement of Low Phase Noise RF and Microwave Sources," *Hewlett-Packard RF & Microwave Meas. Symp.*, April 1979.

[95] D. E. Phillips, "Random Noise in Digital Gates and Dividers," *Proc. 41st Annual Freq. Cont. Symp.*, pp. 507-11, 1987.

[96] W. F. Egan, "Modeling Phase Noise in Frequency Dividers," *IEEE Trans. on Ultrasonic, Ferroelectrics and Freq. Cont.*, vol. 37, no. 4, pp. 307-15, July 1990.

[97] W. F. Egan, "Phase Noise Modeling in Frequency Dividers," *Proc. 45th Annual Symp. on Freq. Cont.*, pp. 629-35, 1991.

[98] M. Bomford, "Selection of Frequency Dividers for Microwave PLL Applications," *Microwave J.*, pp. 159-67, Nov. 1990.

[99] M. R. McClure, "Residual Phase Noise of Digital Frequency Dividers," *Microwave J.*, pp. 124-30, March 1992.

[100] B. Miller and R. J. Conley, "A Multiple Modulator Fractional Divider," *IEEE Trans. on Instrumentation and Measurement*, vol. 40, no. 3, pp. 578-83, June, 1991.

[101] M. Fujishoma, K. Asada, Y. Omura and K. Izumi, "Low-Power 1/2 frequency Dividers Using 0.1-μm CMOS Circuits Built with Ultrathin SIMOX Substrate," *IEEE J. Solid-State Circuits,* vol. 28, no. 4, pp. 510-12, April 1993.

[102] Y. Kado, M. Suzuki, K. Koike, Y. Omura and K. Izumi, "A 1-GHz/0.9-mW CMOS/SIMOX Divide-by-128/129 Dual-Modulus Prescaler Using a Divide-by-2/3 Synchronous Counter," *IEEE J. Solid-State Circuits,* vol. 28, no. 4, pp. 513-17, April 1993.

[103] B. Razavi, K. F. Lee and R. H. Yan, "Design of High-Speed low-Power Frequency Dividers and Phase-Locked Loops in Deep Submicron CMOS," *IEEE J. Solid-State Circuits,* vol. 30, no. 2, pp. 101-9, Feb. 1995.

[104] P. Larsson, "High-Speed Architecture for a Programmable Frequency Divider and a Dual-Modulus Prescaler," *IEEE J. Solid-State Circuits,* vol. 31, no. 5, pp. 744-8, May 1996.

[105] B. Chang, J. Park and W. Kim, "A 1.2 GHz CMOS Dual-Modulus Prescaler Using New Dynamic D-Type Flip-Flops," *IEEE J. Solid-State Circuits,* vol. 31, no. 5, pp. 749-52, May1996.

[106] J. Craninckx and M. Steyaert, "A 1.75-GHz/3-V Dual Modulus Divide-by-128/129 Prescaler in 0.7-μm CMOS," *IEEE J. Solid-State Circuits,* vol. 31, no. 7, pp. 890-7, July 1996.

[107] F. L. Walls, S. R. Stein and J. E. Gray, "Design Considerations in State-of-the-Art Signal Processing and Phase Noise Measurement Systems," *Proc. 30th Annual Freq. Cont. Symp.*, 1976.

[108] T. F. Hock, "Synthesizer Design with Detailed Noise Analysis," *RF Design*, pp. 37-48, July 1993.

[109] T. Kwasniewski, M. Abou-Seido, A. Bouchet, F. Gaussorgues, and J. Zimmerman, "Inductorless Oscillator Design for Personal Communications Devices - A 1.2μm CMOS Process Case Study," *Proc. CICC*, pp. 327-330, May 1995.

[110] L. DeVito, J. Newton, R. Croughwell, J. Bulzacchelli and F. Benkley, "A 52MHz and 155MHz Clock-Recovery PLL," *ISSCC Dig. Tech. Papers*, pp. 142-3, Feb. 1991.

[111] A. W. Buchwald, K. W. Martin, A. K. Oki and K. W. Kobayashi, "A 6-GHz Integrated Phase-Locked Loop Using AlCaAs/Ga/As Heterojuntion Bipolar Transistors," *IEEE J. Solid-State Circuits*, vol. 27, pp. 1752-62, Dec. 1992.

[112] B. Lai and R. C. Walker, "A Monolithic 622Mb/s Clock Extraction Data Retiming Circuit," *ISSCC Dig. Tech. Papers*, pp. 144-4, Feb. 1993.

[113] R. Farjad-Rad, C. K. Yank, M. Horowitz, and T. H. Lee, "A 0.4mm CMOS 10Gb/s 4-PAM Pre-Emphasis Serial Link Transmitter," *Symp. on VLSI Circuits Dig. Tech Papers*, June 1998.

[114] W. D. Llewellyn, M. M. H. Wong, G. W. Tietz and P. A. Tucci, "A 33Mbi/s Data Synchronizing Phase-Locked Loop Circuit," *ISSCC Dig. Tech. Papers*, pp. 12-13, Feb. 1988.

[115] M. Negahban, R. Behrasi, G. Tsang, H. Abouhossein and G. Bouchaya, "A Two-Chip CMOS Read Channel for Hard-Disk Drives," *ISSCC Dig. Tech. Papers*, pp. 216-7, Feb. 1993.

[116] M. G. Johnson and E. L. Hudson, "A Variable Delay Line PLL for CPU-Coprocessor Synchronization," *IEEE J. Solid-State Circuits,* vol. 23, no. 10, pp. 1218-23, Oct. 1988.

[117] I. A. Young, J. K. Greason and K. L. Wong, "A PLL Clock Generator with 5 to 110MHz of Lock Range for Microprocessors," *IEEE J. Solid-State Circuits,* vol. 27, no. 11, pp. 1599-607, Nov. 1992.

[118] J. Alvarez, H. Sanchez, G. Gerosa and R. Countryman, "A Wide-Bandwidth Low-Voltage PLL for PowerPCTM Microprocessors," *IEEE J. Solid-State Circuits,* vol. 30, no. 4, pp. 383-91, April 1995.

[119] I. A. Young, J. K. Greason, J. E. Smith and K. L. Wong, "A PLL Clock Generator with 5 to 110MHz Lock Range for Microprocessors," *ISSCC Dig. Tech. Papers*, pp. 50-1, Feb. 1992.

[120] M. Horowitz, A. Chen, J. Cobrunson, J. Gasbarro, T. Lee, W. Leung, W. Richardson, T. Thrush and Y. Fujii, "PLL Design for a 500Mb/s interface," *ISSCC Dig. Tech. Papers*, pp. 160-1, Feb. 1993.

[121] D. H. Shen, C. M. Hwang, B. B. Lusignan and B. A. Wooley, "A 900-MHz RF Front-End with Integrated Discrete-Time Filtering," *IEEE J. Solid-State Circuits,* vol. 31, no. 12, pp. 1945-54, Dec. 1996.

[122] C. D. Hull, "A Systematic Approach to the Analysis of Noise in Mixers," *IEEE Trans. Circuits and Systems I,* vol.40, no.12, pp. 909-19, Dec. 1993.

[123] W. A. Gardner, *Cyclostationarity in Communications and Signal Processing*, IEEE Press, 1993.

[124] A. L. McWhorter, *Semiconductor Surface Physics*, edited by R. h. Kingston, Philadelphia, PA: University of Pennsylvania Press, 1956.

[125] T. G. M. Kleinpenning, "1/*f* noise in p-n diodes," *Physica*, vol. 98B, pp. 289-299, 1980.

[126] E. A. Leventhal, "Derivation of 1/*f* noise in silicon inversion layers from carrier motion in a surface band," *Solid-State Electronics*, vol. 11, pp. 621-627, 1968.

[127] M. B. Das and J. M. Moore, "Measurements and Interpretation of Low-Frequency Noise in FET's," *IEEE Trans. Elect. Devices*, vol. ED-21, no. 4, April 1974.

[128] K. H. Duh and A. van der Ziel, "Hooge Parameters for Various FET Structures," *IEEE Trans. Elect. Devices*, vol. ED-32, no. 3, March 1985.

[129] L. K. J. Vandamme, X. Li and D. Riguard, "1/*f* Noise in MOS Devices, Mobility or Number Fluctuations?," *IEEE Trans. Elect. Dev.*, vol. 41, no. 11, pp. 1936-45, Nov. 1994.

[130] J. Chang, A. A. Abidi and C. R. Viswanathan, "Flicker Noise in CMOS Transistors from Subthreshold to Strong Inversion at Various Temperatures," *IEEE Trans. Elect. Dev.*, vol. 41, no. 11, pp. 1965-71, Nov. 1994.

[131] I. Bloom and Y. Nemirovsky, "1/*f* noise reduction of metal-oxide-semiconductor transistors by cycling from inversion to accumulation," *Appl. Phys. Lett.* 58 (15), 15 April 1991, pp. 1664-66.

[132] S. L. J. Gierkink, A. van del Wel, G. Hoogzaad, E. A. M. Klumperink and A. J. M. van Tuijl, "Reduction of the 1/*f* Noise Induced Phase Noise in a CMOS Ring Oscillator by Increasing the Amplitude of Oscillation," *Proc. ISCAS '98*, June 1998.

[133] S. L. J. Gierkink, E. A. M. Klumperink, T. J. Ikkink and A. J. M. van Tuijl, "Reduction of Intrinsic 1/*f* Device Noise in a CMOS Ring Oscillator," To appear in *ESSCIRC 98*, The Hague, Sept. 1998.

[134] A. A. Abidi, "High-Frequency Noise Measurements of FET's with Small Dimensions," *IEEE Trans. Elec. Devices*, vol. ED-33, no. 11, pp. 1801-1805, Nov. 1986.

[135] T. Blalack, J. Lau, F. J. R. Clement and B. A. Wooley, "Experimental Results and Modeling of Noise Coupling in a Lightly Doped Substrate," *Tech. Dig. IEDM*, Dec. 1996.

[136] Y. P. Tsividis, *Operation and Modeling of the MOS Transistor*, New York: McGraw-Hill, 1987.

[137] D. DeMaw, *Practical RF Design Manual*, p. 46, Prentice-Hall, 1982.

[138] W. A. Gardner, *Introduction to Random Processes*, McGraw-Hill, Inc., 1990.

[139] A. Papoulis, *Probability, Random Variables, and Stochastic Processes*, McGraw Hill, Inc., 1991.

[140] C. Kittel, *Introduction to Solid State Physics*, John Wiley and Sons, Inc. New York, 1986

[141] *HSPICE User's Manual*, Meta Software, 1996.

[142] *RDL NTS-1000A Phase Noise Measurement System, User's Manual.*

[143] A. Tabatabaei, *private communications.*

Index

D

E

F